Syllogistic Logic and Mathematical Proof

Syllogistic Logic and Mathematical Proof

PAOLO MANCOSU AND
MASSIMO MUGNAI

OXFORD
UNIVERSITY PRESS

Great Clarendon Street, Oxford, OX2 6DP,
United Kingdom

Oxford University Press is a department of the University of Oxford.
It furthers the University's objective of excellence in research, scholarship,
and education by publishing worldwide. Oxford is a registered trade mark of
Oxford University Press in the UK and in certain other countries

Published in the United States of America by Oxford University Press
198 Madison Avenue, New York, NY 10016, United States of America

British Library Cataloguing in Publication Data
Data available

Library of Congress Control Number: 2023933211

ISBN 978–0–19–887692–2

DOI: 10.1093/oso/9780198876922.001.0001

Printed and bound in the UK by
Clays Ltd, Elcograf S.p.A.

Contents

Acknowledgments

While writing this book we have incurred many debts. We have decided to only list in alphabetical order the names of those friends and colleagues who have directly contributed with their comments to our text. We are confident that even though we do not list individually their specific contributions, they know how much we are grateful to each one of them. Thanks to Francesco Ademollo, Lanier Anderson, Paola Basso, Michael Beeson, Wolfgang Carl, Vincenzo De Risi, Daniel di Liscia, Alessandro Giordani, Orna Harari, Desmond Hogan, Eberhard Knobloch, Maria Rosa Massa, Richard McKirahan, Bernardo Mota, John Mumma, Marco Panza, Francesco Paoli, Carl Posy, David Rabouin, Stephen Read, Roshdi Rashed, Paul Rusnock, Boaz Faraday Schuman, Sun-Joo Shin, Riccardo Strobino, Paul Thom, Justin Vlasits, and Richard Zach.

We are also grateful to two anonymous reviewers for Oxford University Press for constructive criticisms that have improved our text. Many thanks also to our editor, Peter Momtchiloff, for having supported our project from the start and for having seen it through publication.

We know that mathematicians care no more for logic than logicians for mathematics. The two eyes of exact sciences are mathematics and logic: the mathematical sect puts out the logical eye, the logical sect puts out the mathematical eye; each believing that it sees better with one eye than with two. The consequences are ludicrous.

Augustus De Morgan
The Athenaeum, July 18, 1868

Introduction

In this volume, we attempt to reconstruct the vicissitudes of a conviction that for centuries prevailed among logicians and philosophers in Western culture. It was the conviction that all theorems of Euclid's *Elements* and, more in general, all mathematical theorems can be proven by means of the traditional Aristotelian syllogism.

As Henry Mendell has shown, Aristotle adopted a twofold attitude toward the syllogism: on the one hand, he had a quite liberal idea of what a syllogistic inference is, attempting to modify and to adapt it to inferences that do not fit its form, while, on the other hand, he claimed that any kind of mathematical theorem can be proven by means of a syllogism in *Barbara*.[1] Otherwise said, Aristotle assumed to be 'syllogisms' inferences that do not have the canonical form of a syllogism but, at the same time, he stated that every valid demonstration ultimately rests on the 'perfect' first figure syllogisms.

This 'duplicity' of Aristotle's attitude toward syllogism is mirrored by the history of logic in the Western tradition: most logicians and philosophers until the second half of the nineteenth century held that the syllogism of Aristotelian origin was the main tool for demonstrating mathematical theorems; to perform their proofs, however, they employed complex arguments that do not have a canonical 'Aristotelian' syllogistic form. A syllogism may appear at some step or other of the demonstration, but in general this latter cannot be considered 'syllogistic' in the traditional, Aristotelian sense.

Only a very small number of authors were aware of this situation, and they reacted in two different ways: some of them simply considered the syllogism as inadequate to prove mathematical theorems, whereas others proposed to expand the classical Aristotelian syllogism adding relations and relational terms to it.

In the second half of the nineteenth century, Augustus De Morgan subordinated the traditional syllogistic to a logic of relations that he had begun to develop. Since De Morgan, the syllogism, thanks to Charles Sanders Peirce and Gottlob Frege, lost some of its peculiar appeal among logicians and

[1] Mendell (1998).

Syllogistic Logic and Mathematical Proof. Paolo Mancosu and Massimo Mugnai, Oxford University Press.
© Paolo Mancosu and Massimo Mugnai 2023. DOI: 10.1093/oso/9780198876922.003.0001

philosophers and became a small part of a more general theory of logic. Thus, our story finds its natural end with De Morgan. Surely, from Aristotle to De Morgan, the span of time is quite broad, but we aim to account for a circumscribed topic, a topic that involves—in Daniel D. Merrill's words—"a very disturbing historical puzzle":

> How could generations of logicians claim that the syllogism is the universal canon of valid deduction, when they could not be sure that it suffices for even the first theorem of Euclid? Or, if they knew that there are some difficulties in doing this, why did they not see the significance of this fact and pursue the matter further?[2]

We cannot claim to provide a complete answer to this puzzle but we hope that our presentation of the major positions characterizing this debate and our interpretations provide at least a partial answer to it. However, some deep questions remain, for us as well as for all other investigators who have addressed the matter, almost as brute facts that do not admit of a satisfactory explanation. For instance, why did no one in the Greek tradition engage in detail with syllogistic reconstructions of mathematical proofs despite Aristotle's claim that all such proofs were reducible to syllogisms?

The relation, and the tension, between logic and mathematics were to be highly significant. For instance, Michael Friedman has claimed that Kant's revolutionary discovery of the synthetic a priori and the postulation of a pure intuition was a consequence of Kant's realization that mathematical proof could not be accounted for using the logic of his time, which Friedman characterized as monadic first-order logic.[3] In this work we would like to go over what we might call the "conditions of possibility" for what Friedman claims to have been Kant's major move in the foundation of his critical philosophy. Indeed, the emergence of a careful analysis of mathematical proofs through the tools of logic was a surprisingly slow process.

As we have said, a very important phase of this process was the discovery of the logic of relations. The first embryonic form of such a logic was developed during the fourteenth century by philosophers like William of Ockham and John Buridan in connection to their discussion of so-called 'oblique inferences,' that is, inferences in which oblique terms occur. The distinction between *oblique* and *right* terms was grammatical and had its roots in the

---

[2] Merrill (1990: 11).   [3] Friedman (1992); we discuss Friedman's interpretation in Section 6.3.

works of the Latin grammarians of antiquity.[4] A *right term* (*terminus rectus*) was simply a term in the nominative case (for example: *Caesar*), whereas an *oblique term* (*terminus obliquus*) was a term in any other case, different from nominative (for example: *Caesaris* 'of Caesar'). Medieval logicians were aware that oblique terms implied a reference to relations and attempted to develop a treatment of oblique terms inside the framework of traditional syllogism.[5] Given their poor interest in mathematics, however, they did not associate oblique inferences with the inferences that were usually carried out by mathematicians when proving theorems.

It is only with the works of Joachim Jungius and Johannes Vagetius, in the seventeenth century, that two important features of oblique inferences emerge: (1) they are for the most part not reducible to syllogisms; (2) they are necessary for proving mathematical theorems. Even though it is quite difficult to determine exactly the influence and the diffusion of Jungius' and Vagetius' theses, there is no doubt that they were a clear symptom of a sort of uneasiness towards the traditional syllogism of Aristotelian origin.

In the seventeenth century, even Gottfried Wilhelm Leibniz and the Spanish philosopher Juan Caramuel Lobkowitz (1606–1682) tackled the problem of oblique (relational) inferences. Leibniz had the opportunity of looking at Jungius' papers and discussed with Vagetius some issues concerning relational inferences. Leibniz was aware of the non-syllogistic nature of certain inferences containing relations; and he believed that they should be 'demonstrated' on the basis of a 'superior' logic, that is, a logic more general than that centered on the syllogism.

Caramuel, by contrast, was an enthusiastic supporter of oblique inferences, which, according to him, constituted the greatest part of our ordinary inferences and attempted even to develop a *logic of oblique terms* (*logica obliqua*). This logic, however, was conceived inside the traditional framework of the doctrine of syllogism, which in the end came out profoundly modified. To integrate relations and relational terms into the traditional syllogistic figures, Caramuel changed the meaning of the ordinary *copula*, thus giving rise to a theory, which strongly resembles the theory of the 'general copula' proposed

---

[4] Varro (1910: 8, 49); Quintilianus (1970: 1.6.25).

[5] In the medieval commentary to Aristotle's *Prior Analytics*, written by Robert Kilwardby (also known as Robertus Anglicus) in the thirteenth century, we find a chapter on oblique inferences (see Kilwardby 2015: 878–91). And while Kilwardby (2015: 885) discusses doubts as to "whether or not it is possible to syllogize from oblique terms" he resolves them by providing arguments to the effect that oblique inferences can be accommodated within the theory of syllogism. The discussion of oblique terms in syllogistic logic finds its roots in Aristotle's treatment of the matter in *Prior Analytics* I.36–I.37.

by De Morgan in the second half of the nineteenth century.[6] Yet, even though Caramuel seems to come very close to developing the beginnings of a logic of relations, he contents himself with amassing heterogeneous examples of 'syllogistic inferences' that contain relations, without making any attempt to elaborate a theory with some degree of generality.

The question of inferences containing relations resurfaces with some force in the second half of the nineteenth century in the United Kingdom, a discussion foreshadowed in the work of the Scottish philosopher Thomas Reid (1710–1796). Reid clearly states that some relational inferences cannot be reduced to syllogisms and that the traditional syllogistic doctrine is unfit to represent mathematical proofs. Reid's statements set off a discussion in which Augustus De Morgan was involved. Whereas Vagetius recognizes the non-syllogistic nature of some relational inferences, insisting at the same time on the necessity of employing them in a mathematical proof *together* with the traditional syllogistic inferences, Thomas Reid concludes from the non-syllogistic nature of those inferences that the syllogism is inadequate to prove mathematical theorems. With De Morgan, however, a logic of relations in a proper sense begins to be built. However, we will show that this was not quite the end of the story and that the conceptual hold of the thesis that all of mathematics could be captured syllogistically was in some cases still exerting its influence in the twentieth century.

Our text is divided into eight chapters and a conclusion. Chapter 1 considers the relationship, such as it was, between logic and mathematics in antiquity and the medieval period. Chapter 2 introduces the treatment of inferences containing oblique terms by, among others, John Buridan and William of Ockham, two of the most eminent logicians active in the fourteenth century. Chapter 3 concerns the emergence in the Renaissance of a more careful analysis of mathematical (Euclidean) proofs with the tools of syllogistic logic. We cover in this section Piccolomini's first detailed syllogization of a geometrical proof as well as similar syllogizations by other Renaissance and seventeenth-century authors. Chapter 4 goes back to the problem of oblique inferences and investigates the treatment of obliquities in several authors active in the seventeenth century. Chapter 5 is devoted to the so-called 'pre-Kantian' philosophy in Germany, including Andreas Rüdiger and Christian Wolff. Chapter 6 deals with Kant. Chapter 7 is devoted to Bolzano and, finally, Chapter 8 is centered on De Morgan's work.

---

[6] Cf., Merrill (1990: 67–71); Dvoràk (2008: 658–9).

Caveat lector. It is well known that defining what exactly is a syllogism is a delicate question that poses deep philological and logical problems (see, e.g., Thom 1981; Smiley 1973; Lear 1986; Corcoran 1972, 1974, and references therein contained).[7] In our case, the complexities are multiplied by the variety of authors who refer to the syllogism in varying degrees of precision. Fortunately, for our purposes we will rarely need to engage in a reconstruction of the precise conception of syllogism entertained by any specific author under discussion (more often than not the authors we discuss give no thorough account of what their conception of a syllogism is). Furthermore, while we have done our best to introduce and explain concepts that might be unfamiliar to the reader, we must presuppose that the reader has already some acquaintance with Aristotelian syllogisms and with Euclid's style of proof in the *Elements.* Finally, our history covers a good number of well-known authors (Aristotle, Leibniz, Kant, Bolzano) but also several lesser-known figures. Lest the reader be confused by the selection, our criteria of inclusion are as follows: we discuss all those authors who have made a contribution to the debate of whether mathematical proofs could be syllogized. Those who did not discuss the topic are not included. Among those who did discuss it, we emphasize the ones who made innovative conceptual moves and give less space to those who simply repeated points that had already been made by other scholars before them. This is the reason why some of the "big and famous" in the history and philosophy of mathematics do not appear and other, in some cases less well-known, characters take center stage.

---

[7] Thom characterizes the most fundamental question about the syllogism as "What is it?" (Thom 1981: 11). For a recent history of conceptions of syllogism from Avicenna to Hegel, see Sgarbi and Cosci (2018). For a general exposition of Aristotle's theory of syllogism, see Crivelli (2012).

1

Aristotelian Syllogism and Mathematics in Antiquity and the Medieval Period

Among the most outstanding achievements of Greek thought were Aristotle's invention of logic as a discipline, as codified in the *Organon*, and the development of proof in mathematics, as exemplified in Euclid's *Elements* and later works. And yet, it comes as somewhat of a surprise that when one looks carefully at the relation between logical demonstration and mathematical proof in the Ancient world (and we should also add the medieval contributions), the results are dismal.

One of the reasons why this is surprising is that there are many programmatic statements in Aristotle's *Organon* that prima facie indicate a close analysis of mathematical demonstrations on the part of Aristotle and his successors. Barnes aptly summarizes the situation:

> In his Elements Euclid first sets down certain primary truths or axioms and then deduces from them a number of secondary truths or theorems. Before ever Euclid wrote, Aristotle had described and commended that rigorous conception of science for which the Elements was to provide a perennial paradigm. All sciences, in Aristotle's view, ought to be presented as axiomatic deductive systems – that is a main message of the *Posterior Analytics*. And the deductions which derive the theorems of any science from its axioms must be syllogisms – that is the main message of the *Prior Analytics*.[1]

Consider for instance what Aristotle tells us in the *Posterior Analytics*:

> The mathematical sciences carry out their demonstrations through this figure [the first figure], e.g. arithmetic, geometry, optics – and in general those sciences which make enquiry about the cause.

---

[1] Barnes (2007: 360).

Syllogistic Logic and Mathematical Proof. Paolo Mancosu and Massimo Mugnai, Oxford University Press.
© Paolo Mancosu and Massimo Mugnai 2023. DOI: 10.1093/oso/9780198876922.003.0002

The puzzle generated by this claim is that it is empirically false. Much of the secondary literature (Barnes, Mendell, etc.) has painstakingly tried to account for such claims but these attempts do not affect the basic fact that strikes the reader of Greek logical texts, namely that almost no attention is devoted to mathematical proof per se. When mathematical examples are brought into play, they are usually instrumental in discussing specific issues that do not question the general assumption that any kind of valid arguments, including mathematical proofs, are, or can be rendered, syllogistically. Typical in this respect is Aristotle's discussion in *Prior Analytics* I. 24 of a mathematical example to the effect that the angles adjacent to the base of an isosceles triangle are equal.[2] The example, in which Aristotle invokes instances of what in Euclid will become a common notion regulating identity,[3] is instrumental in showing that in every proof there must be at least one universal premise. However, no real attention is devoted to the internal structure of the mathematical proof. The example is interesting because both Alexander of Aphrodisias and Philoponus discuss it too.[4] And both use the example to emphasize that the proper syllogistic reconstruction of the mathematical proof requires a universal axiom (concerning identity) while at the same time displaying that, for them, the syllogistic nature of the mathematical proof is a dogma that cannot be questioned.

The situation is the same when one looks at Kilwardby's commentary on *Prior Analytics* where the matter is discussed in *Lectio* 29.[5] Kilwardby is very explicit about the syllogistic steps involved in the reconstruction of the theorem in question:

The thesis to prove, therefore, is that E, F (which are the angles to the base) are equal, in the following way:

---

[2] For uses of mathematics in Aristotle, see Heath (1949) and Mueller (1974).

[3] Namely, equal things remain when equal things are subtracted from equals.

[4] See Alexander of Aphrodisias (2006: 16–17, 47–8) and Philoponus (1905: 41^{b}13 pages 253, line 26, to 254, line 23). We discuss Philoponus in Chapter 3.

[5] Kilwardby (2015: 677–93). The expression "angles of a segment [anguli inscissionis]" is explained by Kilwardby at the beginning of Lectio 29. It refers to the angles C and D in the diagram. The angles of semicircle referred to in the proof are the curvilinear angles that we could denote by CAB and DBA (Kilwardby uses AC and BD, respectively, to indicate them).

All angles of semicircles are equal angles;

AC and BD are angles of semicircles;

So, angles AC and BD are equal.

Let another syllogism be formed as follows:

All angles of a segment [anguli inscissionis] are equal;

But the angles C and D are angles of a segment;

So, they are equal.

Then let a third syllogism be formed as follows:

From all equals, if equals are subtracted, the remainders will be equal;

But the angles AC and BD are equal, likewise the angles C [and] D (as has been shown);

So, when the angles C and D are subtracted from the angles AC and BD, the remainders will be equal.

But the remainders are the angles on the base, viz. E and F; hence the angles on the base are equal.[6]

However, the syllogistic reconstruction, while detailed, only puts in more formal garb and does not question what was assumed all along. The syllogistic proof is used only as a bridge to a lengthy discussion of the claim that a universal premise must be present in every syllogism and consequences thereof. We should point out that Kilwardby's treatment is unusually detailed in presenting the syllogistic formulation of the theorem.

We have been discussing Aristotle's logic theory as centered on syllogisms. But what exactly was included in such theory? Here we have to warn the reader that syllogism is "said in many ways."[7] In Aristotle we find, to start with, a very broad meaning of syllogism according to which a syllogism is simply a deductively valid argument.[8] There is, however, a more restricted notion that is captured by Barnes as follows:

Aristotle's predicative syllogistic is, or can be reconstructed as, an axiomatized deductive system the axioms (or quasi-axioms) of which are two

---

[6] Kilwardby (2015: 680–3).   [7] See our *Caveat lector* at the end of the Introduction.

[8] This very broad meaning of syllogism is found in the definition of syllogism given in *Topics* I.1 and *Prior Analytics* I.1. For instance, in *Prior Analytics* I.1 we read: "A syllogism is a discourse in which, certain things being stated, something other than what is stated follows of necessity from their being so." While quite broad, this definition of syllogism is more restrictive than that of a "logically valid argument" from a contemporary point of view. For instance, circular arguments (i.e., arguments containing a premise identical to the conclusion) or arguments with contradictory premises seem to be excluded by Aristotle as being syllogistic. We thank Francesco Ademollo for an enlightening conversation on this matter.

syllogistic forms, certain principles of conversion and of subordination, a principle of reduction to the impossible, and a rule of exposition or ecthesis. And the theorems (or quasi-theorems) are certain derived principles of conversion and subordination – and an infinite number of syllogisms.[9]

The dogma we have referred to, and which has its roots in Aristotle, is that every syllogism in the broader sense is reducible to a syllogism in the more restricted sense.[10] Even within the more restricted sense, there is quite a bit of freedom in determining what Aristotle accepts as a categorical statement, what type of terms may appear in the syllogism, how the copula 'belonging' ought to be interpreted, and so on. Hence, the alternative attempts to capture exactly what the precise details of the theory are.[11]

One of the few voices[12] that seems to stand in opposition to this state of things is that of Galen who introduces a new class of syllogisms, called *relational syllogisms*. These syllogisms differ, according to him, from categorical and from hypothetical syllogisms:

> There is also another, third, species of syllogism useful for proofs, which I say come about in virtue of something relational, while the Aristotelians are obliged to number them among the predicative syllogisms.[13]

The relational syllogisms are characterized by the fact that a new axiom is needed to account for their validity or, as Galen says, relational syllogisms are 'in accordance with an axiom.'[14] It is possible that Galen is here reacting to Alexander's attempt to reduce the inferential pattern with identity present

---

[9] Barnes (2007: 367).

[10] Aristotle offers a proof of the claim at *Prior Analytics* I.25, 41b36–42a32 and I.23, 40b18–41a21. This thesis is (implicitly or explicitly) opposed by all those thinkers who claim that there are non-syllogistic inferences. We will encounter many of them in our work, including Jungius, Leibniz, Rüdiger, Bolzano, and De Morgan. See the discussion of the Aristotelian argument we give in Chapter 7 on Bolzano.

[11] Mueller (1974) refers in this connection to Aristotle's "vagueness." A classical attempt to come to terms with this "vagueness" is Mendell (1998).

[12] We do not say "the only voice" for one finds in the Megaric and Stoic schools the doctrine of the *lógoi amethódos peraínontes*, which are enthymematic arguments considered to be not reducible to the theory of syllogism. For a presentation of this debate, see Barnes (2012: 104–11).

[13] Galen (1964: xvi 1).

[14] It is possible that this characterization, according to Galen, might have a broader application than just for relational syllogisms. On this topic, see Barnes (2007: 419–47) who seems to conclude that Galen had progressively extended the application of the criterion to claim that all syllogisms depend for their validity on a universal axiom.

in Euclid I.1[15] to the realm of categorical syllogism.[16] Galen emphasized the importance of such relational syllogisms to "arithmeticians and calculators." He also presents a geometrical instance of relational syllogism:

> Given that there is this universal axiom, which has its warranty from itself – namely, items equal to the same item are also equal to one another – it is possible to syllogize and prove in the way in which Euclid produced the proof in his first theorem where he shows that the sides of the triangle are equal; for since items equal to the same item are also equal to one another, and it has been shown that the first and the second are equal to the third, the first will be equal to the second.[17]

Galen stands out in his recognition that something special is going on with inferences that involve relations such as 'equal to,' 'larger than,' and so on. Yet, this attention to the workings of specific mathematical inferential patterns (involving 'identity,' 'greater than,' 'smaller than,' and other relations) is as far as Galen ventures and no logic of relations emerges from his work.

Thus, with the possible exception of Galen, we see no reason to disagree with Mueller who summarizes the relation between Aristotelian syllogistics and mathematical proof in the Greeks as follows:

> Aristotle seems, then, to have had a largely a priori conception of the relation between his logic and mathematical proof. He may have taken the formulation of mathematical theorems into account in trying to justify his estimation of the significance of the categorical proposition in demonstrative science, but his notion of the categorical proposition was so broad that virtually any general statement would satisfy it. On the other hand, Aristotle does not seem to have looked at mathematical proof in any detail, at least as far as its logic is concerned. He recognizes some common features of mathematical

---

[15] We say "possible" because the question of datation concerning Galen's and Alexander's lives is debated; some claim that they were near contemporaries while others claim that Galen lived before Alexander by one or two generations, in which case it is harder to argue that Galen is reacting to Alexander. We will discuss Euclid I.1 at length in Chapter 3. It is the problem of constructing an equilateral triangle on any given segment.

[16] Philoponus is no different than Alexander in giving primacy to categorical syllogism (see Mueller 1974 and Philoponus 1905: 41^{b}13 pages 253, line 26 to 254, line 23). Both Alexander and Philoponus point out that an additional axiom is needed to complete the syllogism but the difference with Galen is that they try to account for the mathematical inference within categorical syllogism. According to Mueller (1974), Stoic reflection on logic seems mostly independent of an analysis of mathematics. However, Posidonius might have been a source for Galen's notion of relational syllogism. See Barnes (2007: 420–33) and Galen (1964: xviii.8).

[17] Galen (1964: xvi.6).

proof, e.g. the use of reductio ad absurdum and the reliance on universal assumptions but he is apparently content to rely on the abstract argument of I.23 to establish the adequacy of syllogistic for mathematics. His peripatetic successors do not seem to have gone much beyond him either in logic or in the logical analysis of mathematical proof.[18]

The situation was not to improve in the medieval period. It is actually not easy to find examples of medieval thinkers (including those active in the Islamic world) who explicitly discuss the issue of whether mathematical proofs can be syllogized.

The most important example we have of such a discussion in the Islamic world comes from Avicenna's *Treatise on Logic*, that is, Book I of the Persian treatise *Philosophy for 'Alâ' al-Dawla* (*Dânishnamah-yi Alâ'î*; written around 1021–37 CE).[19] The example was reproduced almost verbatim in a work by al-Ghazālī (the Latin Algazel).[20] Since the late twelfth century, a work by Algazel circulated in the Latin West, often with the title *Summa theoricae philosophiae* (or, later, *Logica et philosophia Algazelis*). The book, described as "a primer on the Aristotelian tradition,"[21] was translated by Dominicus Gundissalinus in the third quarter of the twelfth century and became well known in the Latin Middle Ages.[22] The *Summa theoricae philosophiae* was a translation of al-Ghazālī's *The Doctrines of the Philosophers* (Maqâṣid al-falâsifa). The text was a loosely adapted Arabic translation of Avicenna's Persian work, *Philosophy for 'Alâ' al-Dawla* (*Dânishnamah-yi Alâ'î*). More precisely, it was a compilation stemming from the parts on logic, metaphysics, and physics in Avicenna's *Philosophy for 'Alâ' al-Dawla* plus additional material from other works by Avicenna.

In the discussion of "compound" syllogisms, Avicenna and Algazel analyze the first proposition of Book I of Euclid's *Elements* and reconstruct its demonstration as structured into four syllogisms. We report the entire section as it appears in Algazel's Logic (omitting the variants):

We have an example of a compound argumentation in the first figure in Euclid. If in facto one wants to construct an equilateral triangle on a given

[18] Mueller (1974: 56–7).
[19] See Avicenna (1971: 36–7); cf., Strobino (2021: 76, note 25), where, in addition to Avicenna, Bahmanyār b. al-Marzubān is also mentioned.
[20] See Lohr (1965: 271–2).   [21] Minnema (2014: 158).
[22] See Griffel (2020) and Janssens (2020).

segment *ab*, and one wants to demonstrate that the triangle is equilateral, one proceeds as follows.

Take the point *a* as center on which one fixes a foot of the compass; then open the compass until the point *b* and draw a circle with center *a*. Similarly, fix a foot of the compass on *b* and then extend the other foot to the point *a* and draw a circle. The two circles are thus equal because they are on the same length [they have equal radius] and intersect on point *g*. From this point draw a segment until point *a*, which yields *ga*. And similarly, from *g* draw another segment until point *a*, which yields *ga*. And similarly, from the point *g* draw another segment until point *b*, which yields *gb*. Then the triangle contained within the three points *abg* is equilateral.

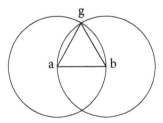

This is proved as follows. The two segments *ab* and *ag* are equal because they originate from the center of the same circle and extend to the circumference of the circle. Similarly the two segments *ab* and *bg* are equal for the same reason. But the segments *ag* and *bg* are equal because they are equal to one and the same segment, namely *ab*. One then concludes that the triangle is equilateral.

One formerly introduced propositions in exactly this way. If, however, they are brought back to their true order, the conclusion will only follow from four syllogisms, each one of which is formed by two propositions:

(1) The first of these syllogisms is this: two segments *ab* and *ag* proceed from the center to the circumference of the same circle. But all the segments that proceed from the center to the circumference of the same circle are equal. Thus the segments *ab* and *ag* are equal.

(2) The second is the following: two segments *ab* and *bg* that extend from the center of the same circumference and extend to the circumference are, for the same reason, also equal.

(3) The third is this: two segments *ag* and *bg* are equal to the same segment, namely *ab*. But any two segments that are equal to the same thing are equal to one another. Thus, the two segments *ag* and *bg* are equal to one another.

(4) The fourth is this: the figure *abg* is contained within three equal segments. But every figure contained within three equal segments is triangular and equilateral. Thus, the figure *abg* constructed on the segment *ab* is an equilateral triangle.

This is the true order of demonstration but the author intended to leave out some propositions because they are obvious. Thus, pay attention to this fact. This is what one needs to say about the form of syllogism.[23]

This is the most explicit syllogistic reconstruction of a Euclidean theorem before Alessandro Piccolomini (1508–1579) provided one in 1547 and we will postpone an analysis of this syllogistic reconstruction of Euclid I.1 until we discuss Piccolomini. As the *Logica Algazelis* was a text well known to the Latin Middle Ages, it is not impossible that it might have influenced Piccolomini. But, surprisingly, it does not seem to have much influenced discussions in the medieval Latin West, for explicit discussions of the possibility of syllogizing Euclid's proofs are difficult to find. Two interesting mentions of syllogized mathematical proofs in the Latin West appear in Albertus Magnus and Robert Kilwardby, both active in the thirteenth century.[24]

We have already mentioned Kilwardby's syllogized presentation of the Aristotelian examples concerning the equality of the angles at the basis of an isosceles triangle. Kilwardby also discusses Aristotle's brief mention of the proof of the incommensurability of the side with the diagonal of a square as an example of syllogism through the impossible (the original passage in *Prior Analytics* is I.23, 41a 26–30). In his commentary of the Aristotelian text, Kilwardby argues, expanding on Aristotle's claim, that if the side and the diagonal of the square were commensurable then it would follow "syllogistically that an even number and an odd number are equal." He then goes on to give a syllogistic reconstruction of the proof.[25] The example is also discussed by Alexander[26] and Philoponus.[27] We will omit the details as they do not add much to what we have already singled out as the attitude toward the "Aristotelian dogma" that is reflected in them.

---

[23] Lohr (1965: 271–2).

[24] We thank Daniel di Liscia and Marco Panza, respectively, for having brought the passages by Albertus Magnus and Robert Kilwardby to our attention.

[25] The proof itself is marred by a mistake made by Kilwardby in establishing one of his premises. However, if the premise is allowed as established, then the syllogistic reconstruction is correct (see Panza 2018).

[26] Alexander (2006: 38–9).   [27] Philoponus (1905: 41^b13 pages 253, line 26 to 254, line 23).

Albertus Magnus touches on the issue of syllogized mathematical proofs in his commentary to Euclid's *Elements*, which has only recently been translated into English. At the end of the proof of Proposition I.1 he writes:

> It may therefore be syllogized thus: Every rectilinear triangle having sides equal to lines going out from the center to the same circumference is equilateral. But triangle ABC has been constituted upon the given line AB having etc.; therefore it is equilateral.[28]

And after providing an alternative proof of the same proposition given by al-Nayrizi he makes a general claim: "it is, moreover, easy to put all demonstrations of this sort into the form of a syllogism."[29] While Albertus' position falls squarely within the traditional assumption that mathematical proofs could be reduced to syllogisms, what is interesting is that it is more difficult to find such claims in commentaries on Euclid's *Elements* than it is to find them in commentaries to the *Prior* and *Posterior Analytics*.

---

[28] Albertus Magnus (2003: 34).

[29] Albertus Magnus (2003: 35). Even Robert Grosseteste, in his *Commentary to the Second Analytics* claims that the first problem of Euclid's *Elements* can be proven by means of five syllogisms (Grosseteste 1981: 95).

2

Extensions of the Syllogism in Medieval Logic

While the search for medieval attempts at syllogizing mathematical proofs yields meager results, there are other aspects of medieval logic that turn out to be important for later reflections on the relation between logic and mathematics. We are referring to the logical contributions on arguments centered on the use of oblique terms and their connection to relational sentences and to the use of expository syllogism. In Section 2.1 we will look at what Buridan, Ockham, and Albert of Saxony say about oblique syllogisms. In Section 2.2 we will analyze the expository syllogism.

2.1 Oblique Terms and Relational Sentences in Late Medieval Logic: John Buridan, William of Ockham, and Albert of Saxony

The distinction between *oblique* and *right* terms is rooted in the works of the Latin grammarians of antiquity.[1] A *right term* (*terminus rectus*) is a term in the nominative case, whereas an *oblique term* (*terminus obliquus*) is a term in any case other than nominative. Thus *homo* ('man') is a right term, whereas *Caesaris* ('of Caesar,' in the genitive), *ensem* ('sword,' in the accusative), *Ciceroni* ('to Cicero,' in the dative case, as, for instance, in 'I will give a book to Cicero') are all oblique terms.

Medieval logicians were well aware that *oblique terms* in many cases hint at relations. So, for example, in Ockham's *Summa Logicae* we read:

> But it suffices to know that for Aristotle every name, or expression having the force of a name (as a participle for instance), which cannot, when significantly taken, be truly predicated of anything unless the oblique form of an

---

[1] See footnote 4 in Chapter 1.

Syllogistic Logic and Mathematical Proof. Paolo Mancosu and Massimo Mugnai, Oxford University Press.
© Paolo Mancosu and Massimo Mugnai 2023. DOI: 10.1093/oso/9780198876922.003.0003

expression [...] is added to it in a true and suitable way, is really relative and falls under the category of relation [...][2]

The introduction of obliquities into categorical sentences permits the construction of premises as 'some man every horse [accusative] is seeing,' 'all bishop's horses are running,' and so on, that, in their turn, give rise to oblique syllogisms (i.e., to syllogisms with at least one premise containing one oblique term).[3] As Paul Thom has remarked, medieval theories of oblique syllogisms "represent a genuine development of the Aristotelian logic, and contain the beginnings of a formal logic of relations."[4]

John Buridan, for instance, discusses at some length the syllogisms containing oblique terms, and the same topic is treated by William of Ockham, Walter Burley, and Albert of Saxony. In the first chapter of the second part of his *Treatise on consequences*, Buridan assimilates the behavior of oblique terms to that of adjectives: both, when associated with a direct term, have the effect of restricting its denotation. According to Buridan, oblique terms, like adjectives, cannot play the role of subjects in a 'simple subject-predicate proposition,' thus revealing a kind of *unsaturatedness* typical of relations and relational terms:

> [...] I believe that just as an adjective, unless it is in the neuter gender of substantives, cannot alone replace the verb nor be the whole subject of a simple subject-predicate proposition, neither can an oblique term.[5]

Even though introducing oblique terms into the syllogism opens the door to the logical treatment of relations, all oblique inferences considered by Buridan in the *Summulae* on syllogism and in the treatise on consequences hold because of the *dictum de omni et nullo* and the traditional theory of distribution. Roughly speaking, the theory of distribution states that a term is distributed if it is either within the scope of a universal quantifier or it is a

---

[2] Ockham (1998: 173). The participle of the verb 'to see,' denoting the proper act of seeing ("significantly taken"), for instance, cannot be "truly" predicated of Socrates (or of any other individual), without specifying what is the object of Socrates' seeing. In other words, 'Socrates is seeing' is an incomplete sentence: to be completed, an oblique expression, as 'a horse' or 'a tree,' which in Latin occurs in the accusative (oblique) case, should be added to the participle, thus giving rise to a complete sentence as 'Socrates is seeing a horse,' 'Socrates is seeing a tree,' and so on.

[3] Parsons (2014: 160–76).

[4] Thom (1977: 143). For a general view concerning the transformations of the syllogism during the Middle Ages, see Thom (2016: 290–315).

[5] Buridan (2015: 128).

predicate in a negative sentence (it is 'undistributed' in the other cases).[6] Buridan presents as follows the *dictum de omni (et nullo)*:

> *Dici de omni* [to be predicated of all] applies when nothing is taken under the subject of which the predicate is not predicated, as in 'Every man runs.' Dici de nullo [to be predicated of none] applies when nothing is taken under the subject of which the predicate is not denied, as in 'No man runs.'[7]

As Van Eijck (1985), van Benthem (1986) and Sánchez Valencia (1991) have pointed out, the *dictum de omni* and the traditional theory of distribution rest on the principle of lexical monotonicity, and this principle assumes two different forms in the case, respectively, of the *dictum* and of the theory of distribution.[8] In Sánchez-Valencia's words:

> On the one hand, upward monotonicity reflects the classical *Dictum De Omni*:
>   'whatever is true of every X is true of what is X.'
> On the other hand, downward monotonicity reflects what traditional logic called distributed occurrence of terms:
>   'a term is distributed in a sentence if the sentence is true about *all of the predicate*.'[9]

Buridan's account of oblique terms, indeed, in perfect agreement with the scholastic and late-scholastic tradition, considers them essentially as modifiers of the direct term to which they are related:

> [...] an oblique term cannot be subjected or predicated in itself, in the sense of standing alone; rather, it is necessary that a nominative be added or implied, to which the oblique term is related as a determination to a determinable.[10]

---

[6] The theory of distribution was criticized by Peter Geach in Geach (1980: 27–48) and Geach (1981: 62–6), but reappraised by Makinson (1969), Van Eijck (1985), van Benthem (1986), Sánchez Valencia (1991), Hodges (1998), and Parsons (2006).

[7] Buridan (2001: 306).

[8] Assuming 'A' and 'B' as names of a natural language and F(A), F(B) sentences containing these names, the rules of, respectively, *upward* and *downward monotonicity* may be represented as follows (Sanchez-Valencia 1991: 23):

Upward Monotonicity	Downward Monotonicity
Every A is B F(A)	Every A is B F(B)
F(B)	F(A)

[9] Sánchez-Valencia (1991: 13).

[10] Buridan (2001: 51). On the same point, cf., Saccheri (1701: 1).

Accordingly, it will first be explained what an oblique term is when it is used with a direct term that is governed by it as a determination of that direct term, just as an adjective is a determination of a substantive [term]. For just as when saying "A white horse is running" the expression "white" determines the expression "horse" to supposit only for white ones, so if I say "Socrates' horse is running", the expression "Socrates's" restricts the expression "horse" only for those that are Socrates's.[11]

Analogous remarks concerning the nature of oblique terms are found in the passage cited above from Ockham's *Summa logicae*. According to Ockham, to the category of relation properly belong those names which to be truly asserted of something need other names in the oblique case to be added to them. As examples of these names, he gives the two pairs 'master-servant,' 'father-son,' and the word 'similar': 'nobody is a father without being father of someone; and nobody is 'similar' without being similar to someone else.'[12]

Thus, introducing oblique terms into the premises of a syllogism, medieval logicians conform themselves to the custom of expanding the Aristotelian syllogism documented since late antiquity. This expansion, in the case of Buridan, involves even the introduction of anaphoric expressions in the syllogism, automatically changing the inference from syllogistic to *non-syllogistic*:

[. . .] because the whole combination is not distributed, it does not follow, "of some human any ass is running, the bishop's ass is [some] human's ass, so the bishop's ass is running," because the bishop's ass is perhaps lying in the stable and Socrates is the human whose every ass is running. So in this case, if we wish to include in a syllogism what is distributed in the proposition in question, we must form a minor premise in which a relative of identity is adjoined to the term "human" so that the middle is understood to hold of the same thing in the premises. For example, it does follow "Of [some] human any ass is running, the bishop's ass is an ass of that human, so the bishop's ass is running."[13]

Buridan's argument may be presented as follows:

(1) Any ass of some human is running;
(2) The bishop's ass is an ass of that human [*eiusdem hominis*];
(3) Therefore, the bishop's ass is running.

---

[11] Buridan (2015: 128).   [12] Ockham (1998: 172).
[13] Buridan (2015: 129). On anaphoric words in medieval logic, cf., Parsons (2014: 227–58).

Clearly, the cross-reference to 'human' in the first and second premise makes the inference non-syllogistic. The same holds for similar 'extensions' of the syllogism, emerging from the following passages:

> Of the king every horse is running, Brownie is a horse of that very king [*eiusdem regis*], so Brownie is running.[14]
>
> [...] for example, "Every horse is running and you have one of those horses, so you have something running."[15]

Buridan is well aware of the non-syllogistic nature of these inferences, and recognizes that they may be considered as belonging to the first figure only in virtue of some kind of *similarity*:

> These syllogisms are as perfect as syllogisms made in the first figure from simple and direct terms. So they can also be reduced to the first figure by their similarity, because they hold straightforwardly and immediately by the nature and definition of distribution and subsumption under a distributed term.[16]
>
> For example, the proposition "A human's ass is running" is equivalent to "Some human is [one] whose ass is running"; so it is possible to syllogize with the one as is done with the other. So in the first figure, or *something like it*,[17] there is a syllogism like this: "Any human whatever's ass is running, Socrates is a human, so Socrates's ass is running" [...][18]

Like Buridan, also Albert of Saxony admits that not all syllogisms containing oblique terms are syllogisms in the proper sense:

> Moreover, I say that these syllogisms, in which not the whole subject but only a part of it is distributed, are not, properly speaking, in any of the three figures, if we want to call the figures according to the way said before. If, however, we want to call first figure a figure where the term distributed in the major is the predicate of the minor premise, and analogously in the cases of other figures, then these syllogisms can properly be put in some of the three figures.[19]

---

[14] Buridan (2015: 132–3).   [15] Buridan (2015: 133).   [16] Buridan (2015: 131–2).
[17] Our emphasis. The original has: "*Et ideo in prima figura, vel ad eius similitudinem.*"
[18] Buridan (2015: 134).   [19] Albert of Saxony (2010: 754).

In the part of the *Summulae* devoted to the syllogism, Buridan shows that he has a quite liberal attitude toward the syllogistic inference. He states, for instance, that the subject and the predicate of a sentence belonging to a syllogism do not need to be simple terms, but that they may be even complex expressions:

> And we should note that although the subject and the predicate are often composed of many words, as in 'Socrates' horse runs swiftly', nevertheless we assume here, as does Aristotle, that the whole subject and the whole predicate are the terms of a proposition, for in making the distinction between the three figures and their modes, just like Aristotle, we do not intend to go here into the further analysis of a proposition beyond that into its subject, predicate and copula.[20]

Buridan holds fast to the traditional analysis of the proposition into *subject* and *predicate* and according to Gyula Klima, it is "this restriction in particular that causes the inherent limitation of Aristotelian syllogistic, positioning it, as it were, midway between propositional logic and quantification theory with respect to their capacity to represent inferential relations between various types of (assertoric, nonmodal) propositions."[21] As Henry Mendell has shown, however, the possibility of considering complex expressions as subject and predicate of a syllogism, recognized even by Aristotle, is the first step toward an expansion of this kind of inference.[22] Buridan, on his part, refers directly to this possibility with the explicit purpose of justifying the unorthodox position of the terms in oblique syllogisms. When introducing oblique syllogisms, indeed, he emphasizes that "we should realize that in syllogisms with oblique or with complex terms, it is not necessary that the syllogistic terms, namely, the middle term and the extremities, be the same as the terms of the premises and the conclusion, namely, their subjects and predicates."[23] Thus, Buridan considers "valid" and "perfect" the syllogism: 'A donkey sees every man; every king is a man; therefore, a donkey sees every king,' of which he gives the following analysis:

> [...] the syllogistic middle term here is 'man', taken in the major in an oblique case and in the minor in the nominative case, which, however, is neither the subject nor the predicate of the major. The minor extremity is

---

[20] Buridan (2001: 306).   [21] Buridan (2001: 306, note 2).   [22] See Mendell (1998).
[23] Buridan (2001: 367).

'king', whereas the major extremity is the remainder, namely, the aggregate of 'donkey' and 'sees', and this aggregate is attributed to the oblique case of 'king' in the same way in the conclusion as it was attributed to the oblique case of 'man' in the major.[24]

So, according to Buridan, in the syllogism "A donkey sees every man; every king is a man; therefore, a donkey sees every king," the expression *donkey sees* plays the role of the *major term*, *king* that of the *minor* and *man* that of the *middle term*. If, as suggested by Klima, we denote as follows: {major}; [middle]; (minor), the terms of this syllogism, we have "A {donkey sees} every [man]; every (king) is a [man]; therefore, a {donkey sees} every (king)." Clearly, if the set of kings is a proper subset of the set of all men and a donkey sees every man, it sees every king as well. Buridan considers this syllogism *similar* to the syllogisms of the first figure, because it obeys to the same principle on which the first figure syllogisms are based, that is the so-called *dictum de omni et nullo*. Concerning the first figure syllogisms, indeed, he writes:

> And I say that the first four modes of the first figure are called 'perfect', for they are evident consequences by virtue of *dici de omni* or *dici de nullo*, as was said in the preceding chapter.[25]

And in chapter 5, introducing oblique syllogisms, he observes:

> There is another way of syllogizing with oblique terms, for whenever in the major proposition some oblique term is distributed, whether affirmatively or negatively, then of whatever term the nominative form of the oblique term is affirmed in the minor, in the conclusion there will be attributed to the oblique form of that term whatever was attributed to the original oblique term in the major. For example, 'Every man's donkey is running [of every man a donkey is running]; every king is a man; therefore, every king's donkey is running [of every king a donkey is running]; similarly, 'No man a donkey sees [no man is seen by a donkey]; every king is a man; therefore, no king a donkey sees [no king is seen by a donkey]'. And this mode of syllogizing is similar to the first figure with nominative terms, for it holds directly on account of the subsumption under a distributed term.[26]

---

[24] Buridan (2001: 367).   [25] Buridan (2001: 322).

[26] Buridan (2001: 366). Commenting on this passage, Klima remarks: "Or, more simply, if the oblique term of the major is distributed, then it can be replaced in the conclusion by the oblique form of the subject of the minor, when the predicate in the minor is the nominative form of the distributed

Thus *Barbara*, for instance, a first figure syllogism "with nominative terms" (i.e., without obliquities), having the form 'Every A is B; every C is A; therefore every C is B,' holds because the term corresponding to 'C' is within the scope of a universal quantifier and 'subsumed' under the term corresponding to 'A,' which in its turn is within the scope of a universal quantifier. In other words, it holds because it obeys the *dictum de omni* principle and the traditional rules of distribution.

In his treatise on *consequences* Buridan points out that some 'oblique' inferences are built according to the following schema:

the first premise states that all items having a given property A have the property B; the second premise states that something, call it *i*, is related to one or more items that are A; the conclusion is that *i* is related to one or more objects having the property B.

These are Buridan's examples:

For example, it follows, 'Every human is running, [it is some] human you see, so [it is] someone running you see'; similarly, it follows, 'Every horse is black, you have a horse in the stable, so you have something black in the stable'; similarly, 'Every ass is running, [it is some] human's [thing that] is an ass, so [it is some] human's [thing that] is running.'[27]

In these cases, too, the conclusion is inferred from the premises according to a principle of lexical monotonicity.

Buridan's conception of oblique inferences is common to William of Ockham, who precedes him, as well as to Albert of Saxony, who wrote after him. Let us briefly continue to pursue what they say about oblique inferences.

In his *Summa logicae*, Ockham too shows awareness that the validity of oblique inferences similar to those mentioned by Buridan rests on the *dictum* and on the principles of distribution:

---

oblique term of the major. This rule in modern 'natural logic' is often referred to as the rule of 'upward monotonicity.'" See Sánchez Valencia (1991); van Benthem and ter Meulen (1985). According to Klima "this rule provides an example of how supposition theory can serve to overcome the limitations of 'pure syllogistic,' which would leave the intrinsic structure of complex terms unanalyzed." Buridan (2001: 366); on the development (and relevance) of supposition theory during the Middle Ages, cf., Parsons (2008: 157–280).

[27] Buridan (2015: 133).

Concerning the syllogism with oblique terms, we need to know that when the major premise concerns the oblique and the minor premise the term in the nominative case, an oblique conclusion always follows; and the syllogism is ruled by the dictum de omni vel de nullo, provided that the obliquity in the conclusion involves the same extreme as that involved in the major premise, so that if the subject of the major premise is an oblique term, it is needed that the subject of the conclusion to be an oblique term as well.[28]

[...] with any proposition concerning a term which is distributed, in nominative or in oblique case, we may have a valid syllogism if we put in the minor premise another term which is subordinated to that term [...] And it seems to me that this conclusion holds in virtue of the nature or the conditions of the distribution; and all this becomes clear as soon as one understands what the name 'distribution' means.[29]

Albert of Saxony, in his *Logic*, comes to the same conclusion. Discussing the syllogism 'A man sees every donkey; Brunellus is a donkey; therefore, a man sees Brunellus,' Albert remarks that all syllogisms similar to this 'hold and are governed by means of the *dictum de omni* or the *dictum de nullo*.'[30] In this case, too, no appeal is made to any rule or principle involving relations.

Buridan considers the syllogism as a kind of *hypothetical proposition*, that is, as a *single* proposition (not as an aggregate of three distinct propositions) "composed of several expressions" connected together by the conjunction 'therefore,' analogous but distinct from a conditional:

I reply that although a syllogism is composed of several expressions, it is nevertheless a single hypothetical proposition, connecting the conclusion with the premises through the conjunction 'therefore'. Further, it can be relegated [reducere] to the species of conditional propositions, for just as a conditional is one consequence, so too is a syllogism, whence a syllogism could be formulated as a conditional, in the following manner: 'If every animal is a substance, and every man is an animal, then every man is a substance'. Strictly speaking, however, a syllogism has an additional feature in comparison to a conditional in that a syllogism posits the premises assertively, whereas a conditional does not assert them.[31]

---

[28] Ockham (1974: 385).   [29] Ockham (1974: 385).   [30] Albert of Saxony (2010: 756).
[31] Buridan (2001: 309).

Syllogisms, for Buridan, are subsumed under the more general concept of a *consequence*. Thus, it is no wonder that, when describing how to transform into perfect syllogisms the imperfect (and valid) syllogisms belonging to figures different from the first figure, Buridan explicitly appeals to non-syllogistic consequences:

> But where the conversion of a premise or of premises is concerned, we should say that if the premises of an imperfect syllogism can be converted into the premises of a perfect syllogism, it follows of necessity that the imperfect syllogism in question is a valid consequence for drawing the conclusion that the perfect syllogism would reach. And this follows by the same principle as before, namely, that *whatever follows from the consequent follows from the antecedent*; for in the aforementioned case the premises of the perfect syllogism follow from the premises of the imperfect syllogism, whence it is necessary that the conclusion that followed from the perfect syllogism follows also from the premises of the imperfect syllogism.[32]

Finally, by including in the syllogism, as we have seen, pronouns and other expressions containing anaphoric references, Buridan realizes a further, very promising expansion of it.

Buridan, however, his propensity to expand the Aristotelian syllogism notwithstanding, reveals a quite conservative attitude toward demonstration, continuing to pay homage to the traditional syllogism and to consider it as the cornerstone on which all logical proof is based. This was surely due to some sort of deference to Aristotle's logical works, but even to Buridan's firm belief that, as we have seen, all inferences that he considered as 'deviant syllogisms' maintain a kind of similarity with the proper, traditional syllogism from which they derive. The deviance, in these cases, was not so conspicuous to Buridan's eyes to undermine the traditional point of view that attributed a central role to syllogism in the logical theory. According to Buridan, indeed, a demonstration must satisfy the following conditions:

> [...] first, [...] it has to contain the subject, the attribute, and the middle term. [...] Second, it has to contain a conclusion consisting of the subject

---

[32] Buridan (2001: 323). Cf., also p. 322: "I say first that, as far as the conversion of the conclusion is concerned, if the conclusion of a perfect syllogism can be converted by a valid consequence, then it is necessary that the converted proposition follows from the same premises from which the conclusion that was converted into it followed, by the principle that *whatever follows from the consequent follows from the antecedent*."

and the attribute [...] Third, it has to contain two premises, one consisting of the middle and the attribute and another consisting of the subject and the middle. [...] Fourth, these premises have to be arranged in a syllogistic figure and in the appropriate mode, so that the conclusion should necessarily follow from the premises.[33]

In other words: "every demonstration has to be a syllogism."[34]

2.2 Expository Syllogism: Identity and Singular Terms

Besides the inclusion of relational terms by means of 'obliquities' into the traditional Aristotelian syllogism, a further important extension of syllogistic accomplished by medieval logicians was the recourse to the so-called *expository syllogism*. The expository syllogism originated with Aristotle's *ekthesis*, that is, roughly speaking, with the procedure of instantiating individuals (or species, depending on a different interpretation) to exemplify some property in a proof. This is a passage of the *Prior Analytics*, in which Aristotle employs *ekthesis*:

> Now, if A belongs to none of the Bs, then neither will B belong to any of the As. For if it does belong to some (for instance to C), it will not be true that A belongs to none of the Bs, since C is one of the Bs.[35]

There is no agreement among Aristotle's commentators concerning the role of *ekthesis* in the syllogistic, that is, whether it is a (relatively) independent part of the machinery of syllogism or an integral part of it.[36] Controversial is even the nature of the 'things' that are exemplified: whether they are individuals or general terms.[37] The procedure of *ekthesis* (in case of an affirmative particular sentence) may be represented as follows:

---

[33] Buridan (2001: 673).   [34] Buridan (2001: 674).

[35] Aristotle (1989: 2–3); *An. Pr.* I.2, 25a: 15–19 (other examples in I. 6, 28a: 23–6; 28b: 20–1).

[36] For a presentation of the mechanism of *ekthesis* in Aristotle's syllogistic and the various interpretations of it, see Thom (1976), Thom (1981: 164–76), and Smith (1982). About the issue of the independence of exposition and expository syllogism in Aristotle's system, we find quite compelling the reasons advanced by Parsons (2014: 41–2) in favor of its *indispensability*.

[37] Jaakko Hintikka considers the Aristotelian *ekthesis* as analogous of the method of instantiation in contemporary logic: cf., Hintikka (1978); for Parsons (2014: 23) it is "a kind of existential instantiation."

Some As are Bs;
c is A;
Therefore, c is B.

As we have remarked above, a problem arises with the interpretation of 'c': is it the name of an individual or of a term (of a species or of a general concept)? Ockham and Buridan interpreted it as name for individuals and they did the same in the case of the *expository syllogism*.

The *expository syllogism* is a kind of 'converse' of *ekthesis* and may be conceived as an analogue of existential generalization. Applied to a positive sentence it can be represented as follows:

c is A
c is B
Therefore, some A is B.[38]

In Ockham's words:

> We need to know that expository is such a syllogism consisting of two singular premises placed in the third figure, which, however, can infer a conclusion either singular or particular or indefinite but not universal [...][39]

During the Middle Ages, the expository syllogism became quite popular and, as has been recently documented by Miroslav Hanke, logicians influenced by the scholastic milieu employed and discussed it even until the second half of the seventeenth century.[40]

In his *Questions* on Aristotle's *Prior Analytics*, Buridan introduces the expository syllogism as follows:

> It is to be noted that a syllogism is called 'expository' in which the middle is a discrete, i.e. singular, term. And this is so precisely because by means of a discrete term, a particular or indefinite proposition is exposed so that it is true. Thus, if you say 'some man runs', and you are asked 'who is that?', and in reply or explaining, it will be said 'that is Socrates', or 'this man' &c.[41]

---

[38] Cf., Parsons (2014: 24).   [39] Ockham (1974: 403).   [40] Hanke (2020).
[41] Buridan, *Quaestiones in Analytica priora*, Liber 1, q. 6 (translation by E. D. Buckner: *The Logic Museum*).

Then he explains:

> Next, concerning the affirmative expository syllogism (the only syllogism we
> are considering in the present) it is to be said that the affirmative expository
> syllogism holds thanks to the rule 'whatever things are the same as one and
> the same thing in number, are mutually the same'. For example, if I say
> 'Socrates is white and the same Socrates is sitting', it follows that who is
> sitting is white, because the premises inform us that a white thing as much as
> a sitting thing is the same as Socrates. Therefore, we must conclude that a
> white thing is the same as a sitting thing.[42]

In the fifth chapter of his *Summulae*, when dealing with the rules for discrim-
inating the valid syllogisms from the invalid ones, Buridan claims that the rule
according to which from two particulars, indefinite or singular premises
nothing follows (because for yielding a syllogism one of the premises has to
be universal) is false. Indeed, he remarks, the expository syllogism "can very
well result from two singulars"; therefore, the rule should be formulated
without specifying 'from singulars.'[43] Then, Buridan states that "every affirmative syllogism holds by virtue of the principle 'whatever things are said to be
numerically identical with one and the same thing, are also said to be identical
between themselves.'"[44] As an example of application of the rule, he gives this:

> [. . .] if a white thing is identical with Socrates and a running thing also is identical
> with Socrates, then it is necessary that a white thing and a running thing should be
> identical; since, therefore, it amounts to the same thing to say 'Socrates is identical
> with a white thing' and to say 'Socrates is white', the inference 'Socrates is white
> and he runs; therefore a running thing is white' is valid.[45]

In case of negative syllogisms, instead, Buridan states that

> they all are valid by virtue of that other principle, namely: 'whatever things
> are so related that one of them is said to be identical and the other is said to
> be not identical with one and numerically the same thing, they necessarily
> have to be said not to be identical with each other'.[46]

---

[42] Buridan, *Quaestiones in Analytica priora*, Liber 1, q. 6.   [43] Buridan (2001: 313).
[44] As Knuuttila (2010: 222) remarks, something analogous to this principle "was also known from
Aristotle's *Sophistici elenchi* (6, 168b: 31–5) as well from Euclid' s *Elements* I, common notion I, and
some other sources."
[45] Buridan (2001: 313).   [46] Buridan (2001: 315).

As an example of negative syllogism, Buridan gives "if A is identical with Socrates and B is not identical with Socrates, then it necessarily follows that A and B are not identical."[47] According to Simo Knuuttila, the two identity principles just mentioned "could be characterized as the rules which directly specify expository syllogisms."[48]

The inclusion of the expository syllogism into the body of medieval logic constitutes a further expansion of traditional syllogism insofar as it acknowledges as syllogistic some inferences involving identities. Moreover, this inclusion licenses the systematic use of sentences with singular terms into the syllogistic machinery.

Singular sentences hold a quite odd position inside traditional logic. The latter recognized as basic for the syllogism the four categorical sentences: universal affirmative: 'Every A is B'; universal negative: 'No A is B'; particular affirmative: 'Some A are B'; particular negative: 'Some A are not B.' Even though the canonical analysis of the sentence into *subject* and *predicate* was probably inspired by the structure of singular sentences, these were not officially admitted among the sentences on which the syllogism rests. Thus, the problem arose how to classify the singular sentences as regards the canonical categorical sentences. Some authors considered them as logically equivalent to universal sentences: speaking of Socrates, indeed, because there is only one, is the same as speaking of 'every' Socrates. Other authors, instead, considered the singular sentences as equivalent to the particular ones and, finally, some others equated them to both, universal *and* particular sentences.[49] Syllogisms with singular sentences are rare in Aristotle's logical writings and this, in the second half of the past century in particular, has sparked a debate concerning Aristotle's attitude about this kind of sentences.[50]

During the Middle Ages, singular sentences were widely accepted as premises of syllogisms, and in the period of the so-called second scholastic (sixteenth and seventeenth centuries), even syllogisms entirely composed of singular sentences were discussed in several logic textbooks. As Miroslav Hanke has shown, for instance, Hurtado de Mendoza in his logic accepts syllogisms with singular terms.[51] Hurtado is well aware that these kinds of syllogisms are not recognized by the Aristotelian tradition, and remarks that

---

[47] Buridan (2001: 315).   [48] Knuuttila (2010: 222).
[49] Leibniz (1965, 7: 211), for example; but in Leibniz (2020: 112–13) singular sentences are considered equivalent to the particular ones.
[50] Cf., Ross (1949: 289), Lukasiewicz (1951: 6–7), Bird (1964: 90), Patzig (1968: 5), Englebretsen (1980), Thom (1981: 78–80).
[51] Hanke (2020: 1).

Albert the Great denies that a syllogism can be composed of singular propositions.[52] He states, however, that Albert's denial concerns "the syllogism aiming at science, which usually does not care about singular things."[53] But in the domain of "created things," a syllogism composed of singular statements may be accepted:

> [...] it cannot be denied that if by means of two propositions we show that two singular terms are identical with a third singular term, then we can infer via an identity (in the domain of created things) that the two extremes are mutually identical.[54]

As an example of a valid syllogism with singular premises Hurtado gives the following:

> Philip's first-born child is the king of Spain
> Alphonse is Philip's first-born child
> Therefore, Alphonse is the King of Spain.[55]

Hurtado remarks that this inference is just as legitimate as the syllogism *Barbara*, but that we don't find it in Aristotle's writings and that this is the reason why inferences like these belong, in some sense, to a 'lesser rank' (*sunt aliquo modo diminutae*).[56]

According to Honoré Fabri, who composed a logic textbook that was published in the year 1646 by the physician Pierre Mounyer, an expository syllogism is a syllogism in which one or more propositions are singular and "in which from a singular proposition a particular one follows that maintains the same quality and quantity of the original terms."[57] Fabri's example is:

> Peter is white
> Peter is a man
> Therefore, some man is white.[58]

Thus, expository syllogism, oblique and singular terms, together with oblique and singular sentences, expanded considerably the traditional doctrine of the Aristotelian syllogism. As Terence Parsons aptly remarks, it is "often suggested

---

[52] Hurtado (1619: 177–9).   [53] Hurtado (1619: 177–9).   [54] Hurtado (1619: 177).
[55] Hurtado (1619: 177).   [56] Hurtado (1619: 177).   [57] Fabri (1646: 310).
[58] Fabri (1646: 310). Fabri gives the example in the form of a conditional: "*si Petrus est albus, cum Petrus sit homo, aliquis homo est albus.*"

that forms of Aristotelian logic can be symbolized in the monadic predicate calculus," but the expansions just discussed "require the use of a slightly richer notion; they are equivalent to forms in monadic predicate logic *with identity*."[59] This notwithstanding, however, and even though not only the theory of the syllogism but even the entire logic (including the theory of *consequences*) during the Middles Ages became an articulated and deeply developed discipline, medieval logicians, as far as we know, usually abstained from employing the syllogism to demonstrate mathematical (geometrical) theorems.[60] In particular, nobody raised the problem of the suitability of the logical apparatus to account for proofs in mathematics. Nor were there mathematicians who attempted to systematically employ the syllogism in their demonstrations.

[59] Parsons (2014: 63, note 63).
[60] "As far as we know": this clause is necessary because many medieval manuscripts, both in logic and mathematics, are still unpublished and/or poorly studied.

3

Syllogistic and Mathematics

The Case of Piccolomini

In antiquity, two different traditions of proof were developed on relatively autonomous grounds:

(1) the merely 'logical' tradition, based on the Aristotelian doctrines of the *Organon* and on the works of thinkers belonging to the Megaric-Stoic school;

(2) the mathematical (geometrical) tradition, mainly represented by Euclid and his *Elements*.

During the Middle Ages, mathematics and logic, according to the Boethian division of knowledge, were taught in different courses of study: logic in the so-called *trivium* and mathematics (arithmetic and geometry) in the *quadrivium*. Logic was considered mainly as a science of language and was by and large investigated separately from mathematics.[1]

In the Middles Ages, logic began to flourish with Abelard in the twelfth century and, approximately at the same time, Adelard of Bath translated into Latin an Arabic manuscript of the *Elements*. Adelard's edition was followed by other translations made from Arabic sources and by the translation of Ishā q ibn Ḥunain's version made by Gerardo da Cremona. In the thirteenth century, Campano da Novara prepared a text of Euclid's work, which was later employed by Luca Pacioli and Niccolò Tartaglia for their editions.[2]

In the second half of the sixteenth century, the *Elements* were translated into modern languages: into Italian (1543), German (1562), French (1564), English (1570), Spanish (1576), and (later) Dutch (1606). In the same period, Greek manuscripts began to circulate in Europe and Bartolomeo Zamberti translated

---

[1] It is true that toward the middle of the fourteenth century, one witnesses an increasing use of mathematical arguments in theology and natural philosophy (for instance, in the 'calculatores'; see, among other publications, Murdoch 1969, Murdoch 1978, and Sylla 1973) but this did not radically alter the distinction between two traditions of proofs.

[2] See Folkerts (1989).

Syllogistic Logic and Mathematical Proof. Paolo Mancosu and Massimo Mugnai, Oxford University Press.

the *Elements* into Latin directly from the Greek (1505), thus opening the path to a new era in the philological reception of the text. In 1533 Simon Grynaeus published the *editio princeps* of the Greek text. In this sense we are authorized to speak of a proper 'rediscovery' of Euclid in the Western World.[3] One should also mention the rediscovery of Proclus' Commentary on the first book of Euclid's *Elements* as an important aspect of this development.

As is well known, humanists and Renaissance thinkers fiercely reacted against scholasticism and, in particular, against scholastic logic; and even though this reaction did not pose an end to the teaching of traditional logic, from the first half of the fifteenth century onward, thanks to the 'rediscovery' of Euclid's books and the publication of works of authors of late antiquity like Heron of Alexandria and Pappus, the standard of rigorous reasoning began to be identified with Euclid's *Elements*, not with the Aristotelian *Organon*. Jacques Pelletier (1517–1582), for instance, introducing his Latin translation of the first six books of the *Elements* (1557) wrote:

> The Dialecticians call 'proof' a syllogism which causes knowledge, i.e. one that concludes from proven premises; and this originates from Geometry. Better: every proof which leads us to truth is geometrical in character. As is very truly said, we are not able to distinguish the true from the false if we have not previously been well acquainted with Euclid.[4]

In the logic books of this period, it was not uncommon to find as an explicit aim the development of the old-fashioned logical matters according to the rigorous method followed by mathematicians. Girolamo Savonarola (1452–1498), for example, introducing his treatise on logic[5] written in 1484 and first published in 1492, remarks that many students "withdraw from the necessary study of logic" either because of the obscurity of Aristotle's books (*propter librorum Aristotelis obscuritatem*) or because of the quantity and variety of sophistries and verbal intricacies afflicting the teaching of the discipline, and states that he aims to offer a synthesis of the entire dialectic "according to the usual practice of the mathematicians." Savonarola simply aligns himself with the general contempt for the scholastic logic that was in vogue among the 'humanists,' and even the appeal to the 'practice of the mathematicians' was in agreement with the 'spirit of the time': as a matter of fact, however, his treatise is quite conventional and does not reflect the "usual practice of the mathematicians."

---

[3] Cf., De Risi (2016).   [4] Pelletier (1557: 12).   [5] See Savonarola (1982).

In 1697, Gerolamo Saccheri, in the dedicatory letter of his *Logica demon-strativa* will explain that the title of this book alludes to "that rigorous method of demonstrating," typical of geometry, "that saves only the first principles and does not accept anything obscure and not evident."[6] Two centuries after the publication of Savonarola's *Logic*, mathematics (geometry) continues to be considered as the paradigm of any demonstration.

As soon as Euclid's *Elements* began to be considered the cornerstone of any kind of logical inference, thus putting into question the central place held until the end of the fourteenth century by Aristotle's *Organon*, some authors like Piccolomini, Herlin, and Dasypodius raised the problem of whether tradi-tional syllogistic was able to demonstrate geometrical theorems.

It is only with Piccolomini's *Commentary on the certainty of mathematics* (1547) that for the first time in the history of Western thought we are offered a careful reconstruction in syllogistic terms of a Euclidean proposition moti-vated by a specific philosophical program. Piccolomini's decision to embark on an analysis of the (supposedly) syllogistic structure of Euclidean proofs is quite interesting. Indeed, he is motivated by the idea of showing that mathe-matics is not an Aristotelian science. An Aristotelian science is made up of scientific syllogisms. Among the features of scientific syllogisms is that they are causal.[7] Piccolomini does not criticize the notion of an Aristotelian science and the role of scientific syllogisms within it. Rather, his goal was to show that the syllogistic reconstruction of geometrical theorems such as Euclid's *Elements* I.1 (see Section 3.1) shows that the syllogisms are not causal contrary to Aristotle's requirements on scientific knowledge. We have thus an implemen-tation of a logical Aristotelian program (showing that syllogisms can capture the logic of mathematics) aimed at undermining the claim that mathematics satisfy the Aristotelian conception of science. It is the double nature of this project that sets Piccolomini apart from his predecessors. This contribution by Piccolomini was to spark an important debate called the *Quaestio de certitu-dine mathematicarum*, that is, the question of whether the certainty of math-ematics was to be accounted for by the causal nature of its proofs or by some other factors. Mancosu (1996) analyzes the immense ramifications of this debate during the seventeenth century but here we only want to emphasize that the extensive interest in the late Renaissance and the seventeenth century on syllogizations of Euclid's proofs, which we will document, is fueled in great

---

[6] Saccheri (1697: unnumbered).

[7] See Mancosu (1996) where the role of the rediscovery of Proclus' Commentary on the first book of Euclid's *Elements* in Piccolomini's program is also emphasized.

part by this debate. While Piccolomini did not question the possibility of reconstructing mathematical theorems syllogistically, his attention to the issue of causality forced him to be extremely careful in the reconstruction. In Section 3.1 we discuss Piccolomini's syllogistic reconstruction of Euclid I.1 and similar attempts in the seventeenth century. In Section 3.2 we analyze whether Piccolomini's reconstruction can be accepted as a successful syllogization of Euclid I.1.

3.1 Piccolomini's Syllogistic Reconstruction of Euclid's *Elements* I.1

Piccolomini's discussion takes place within the context of Renaissance debates on method,[8] and especially the extensive discussions that occupied the Paduan school on so-called compositive (synthetic) and resolutive (analytic) methods. In chapter 10 of his treatise, Piccolomini explains the distinction. He tells us that Euclid, most of the time, proceeds synthetically. When attempting to grasp the synthetic procedure it can be of great help to proceed *à rebours*, that is, from the conclusion to the premises of the conclusion, from there to the premises of the premises, and so on. That is the resolutive method:

> Indeed, when in trying to grasp such a compositional procedure we move backwards from the last conclusion, passing through the premises and the premises of the premises, until we reach the first principles, we say that we are resolving [analyzing]. And with this resolution we come to know whether that conclusion was in the first place made up from true and adequate premises. Proclus says this in several places, in the first and second books [of the Commentary] on Euclid's first book. Averroes makes the same claim in the first book on the Posterior Analytics and so does Alexander in the introduction to the Prior Analytics. Philoponus states the same in the most evident way in the first book on the Posterior Analytics, chapter 9, where he uses the first problem in Euclid in order to show the entire Euclidean procedure of compositions and resolutions.[9]

---

[8] See Mancosu (1996), Freguglia (1999), and Cozzoli (2007) for extensive references to the mathematical and the philosophical context of Piccolomini's contribution. See also Pedrazzi (1974). For an overview of the relation between logic and mathematics in the seventeenth century, see Mugnai (2010).

[9] Piccolomini (1547: 98).

When we look at Philoponus' commentary on chapter 9 of Book I of *Posterior Analytics* we find a discussion of I.32 that can plausibly be taken to exemplify what Piccolomini ascribed to Philoponus.

However, it is likely that Piccolomini gave the wrong section of the *Posterior Analytics*. Indeed, there is a discussion of Euclid I.1 in Philoponus' commentary on the *Posterior Analytics*, Book I, chapter 8, which strongly hints at the possibility of syllogizing Euclid's propositions, and in particular of Euclid's *Elements* I.1. Philoponus says:

> However, the minor [premises] are contained potentially in the major premises, namely, the axioms. For example, if three straight lines A, B, and C are posited, if I say that since A and B are both equal to C, A and B are therefore equal to one another because 'things which are equal to the same thing' are 'also equal to one another', clearly the minor premise, 'A and B are equal to C' is contained in the [premise] that says 'things which are equal to the same thing are also equal to one another'. And in the deduction 'man is an animal, animal is a substance, man is a substance', the minor premise, 'man is an animal', is clearly included in the major [premise], 'animal is a substance'. And so, there are three things that should be assumed in advance universally in every demonstration: the given, the sought, and the axiom. For example, in Euclid's first theorem, which investigates constructing an equilateral triangle on a given finite straight line, the finite straight line is the given, the equilateral triangle is the sought, and the axiom in the preliminary deductions is 'straight lines' 'extending' from the centre 'to the circumference of a circle' 'are equal to one another' and that 'things which are equal to the same thing are also equal to one another' and [the axiom] in the conclusion is that a triangle contained by three equal straight lines is equilateral.[10]

The translator of this section of Philoponus' commentary, Richard McKirahan, has no doubt that Philoponus in the passage just quoted is thinking of a syllogized proof of Euclid I.1:

> The sequel of this passage makes it clear that Philoponus is thinking of a syllogized version of geometrical proofs, where some of the steps consist in applying a general principle, such as 'things which are equal to the same thing are also equal to one another', to particular cases, for example, the straight lines A, B, and C. A, B, and C are an arbitrary set of lines that

---

[10] Philoponus (2008: 21).

instantiate the principle, and there is an infinite number of other lines that would instantiate it just as well.[11]

We have cited this passage from Philoponus, for it shows the highest level of detail, such as it was, devoted by Greek sources to the problem of syllogizing Euclid. We will now see that Piccolomini's analysis goes further than any of his Greek and Latin sources.

Let us quickly recall Euclid's proposition I.1. First of all, it is important to point out that Euclid I.1 is properly a *problem* and not a *theorem*. This is relevant to our account, for when it comes to problems, the syllogistic account concerns only one part of Euclid's entire text, namely the *apodeixis*. As is well known, Proclus distinguishes six parts in Euclidean propositions. The Greek terms are: *protasis, ekthesis, diorismos, kataskeuē, apodeixis, sumperasma*.[12]

Here is the proof of Euclid I.1 as presented by Mueller following Proclus' structure.[13]

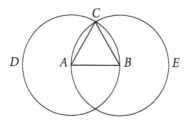

Protasis On a given finite straight line to construct an equilateral triangle.

Ekthesis Let *AB* [...] be the given finite straight line.

Diorismos Thus it is required to construct an equilateral triangle on the straight line *AB*.

Kataskeuē With center *A* and distance *AB* let the circle *BCD* have been described; again with center *B* and distance *BA* let the circle *ACE* have been described; and from the point *C* in which the circles cut one another to the points *A*, *B* let the straight lines *CA*, *CB* have been joined.

Apodeixis Now since the point *A* is the center of the circle *CDB*, *AC* is equal to *AB*. Again, since the point *B* is the center of the circle *CAE*, *BC* is equal to *BA*. But *CA* was also proved equal to *AB*; therefore, each of the straight lines *CA*, *CB* is equal to *AB*. And things which are equal to the same thing are equal to one another; therefore, *CA* is also equal to *CB*. Therefore, the three straight lines *CA*, *AB*, *BC* are equal to one another.

Sumperasma Therefore the triangle *ABC* is equilateral; and it has been constructed on the given finite straight line. Which was required to be done (Q.E.F.).

---

[11] Philoponus (2008: 118).

[12] Here is the passage in Proclus, according to Morrow's translation Proclus (1992 [1970]: 159): "every problem and every theorem that is furnished with all its parts should contain the following elements: an enunciation, an exposition [sometimes translated as setting-out], a specification [sometimes translated as definition of goal], a construction, a proof, and a conclusion."

[13] Mueller (1981: 11).

In the case of theorems, the structure is slightly modified but what is essential here is that the syllogistic reconstruction concerns only the *apodeixis* and not the construction (*kataskeuē*).[14] Piccolomini begins the syllogistic reconstruction starting from the conclusion. His summary of Euclid's *apodeixis* is identical to the *apodeixis* given above by Mueller with the exception of some of the ordering of the letters in the labeling of circles and segments (for instance, circle ACE in Piccolomini corresponds to circle CAE in Mueller's reconstruction).[15]

Piccolomini claims that analysis (*resolutio*) reveals that four syllogysms have been employed in the *apodeixis*.[16] Starting from the conclusion he summarizes the first syllogism (but fourth in the order of composition) as:

First Syllogism
Every figure contained within three equal lines is an equilateral triangle;
The figure ABC is contained within three equal lines AB, AC, [BC]
Thus, the figure ABC is an equilateral triangle.[17]

Piccolomini takes for granted that this is a syllogism and worries only about the justification that can be given of the major and the minor premises. Concerning the major premise, he says that it is evident on account of the definition of equilateral triangle. The minor is justified as a conclusion from another syllogism (the second in the order of resolution but the third in the order of composition), namely

Second Syllogism
Things that are equal to a third are equal to one another;
But AC and BC are equal to the third line AB;
Thus, AC and BC are equal to one another.[18]

---

[14] Mueller (1981: 11).

[15] Piccolomini calls the entire *apodeixis* "ostensio." Clavius also speaks of the proof as an *ostensio* (exhibition) that the construction is correctly set up [*institutam*]: "In omni problemate duo potissimum sunt consideranda, constructio illius, quod proponitur, et demonstratio, qua ostenditur, constructionem recte esse institutam…" Clavius (1612: 28).

[16] The four syllogisms are essentially the same as those found in Algazel's *Logic*, which we discussed in Chapter 1. Algazel's *Summa theoricae philosophiae* appeared for the first time in print in 1506 with the title *Logica et philosophia Algazelis Arabis*, ed. Peter Liechtenstein (Venice, 1506; repr., Frankfurt, 1969). It was reprinted at Venice under the same title in 1536 (repr., Hildesheim, 2001). We have not been able to determine whether Piccolomini is relying on this text or whether he arrived at his analysis independently, although we strongly suspect the first alternative to be the case; however, for our purposes nothing much hinges on this.

[17] Piccolomini (1547: 101a, chapter X). The labeling of the diagram used by Piccolomini is the same as the labeling given in the previous diagram when discussing Mueller's presentation of Euclid I.1.

[18] Piccolomini (1547: 101b).

Once again, no effort is spent in trying to argue that this is a syllogism. Piccolomini claims that the major premise is evident to the mind (as a primary truth). The minor premise is analyzed as a conjunction, for Piccolomini says "the minor premise is proved in relation to each of its parts" and proceeds to show first that BC is equal to AB and then that AC is equal to AB. Each statement is justified by a syllogism. In both cases it is taken for granted that the arguments presented are syllogistic and Piccolomini worries only about the justification of the premises. The third syllogism is:

Third Syllogism
All the lines from the center of the circle to the circumference are equal.
But BC and AB connect the center of the circle ACE to the circumference.
Hence, BC and AB are equal.[19]

The major premise is justified in force of definition 15 in Euclid.[20] The minor by construction. The fourth and last syllogism is analogous, and the premises are justified exactly as those of the third syllogism:

Fourth Syllogism
All the lines that connect the center of the circle to the circumference are equal.
But AC and AB connect the center of the circle BCD to the circumference;
Hence, AC is equal to AB.[21]

Piccolomini triumphantly concluded: "This is a complete resolution, for we have reached immediate and indemonstrable things and nothing can be further resolved."[22]

Before we move to the detailed discussion of the syllogistic reconstruction provided by Piccolomini we point out that it was rather influential. Indeed, even though Piccolomini is not quoted, Clavius in his edition of Euclid's *Elements* (*Scholium* to proposition I.1 in Clavius 1591) provides a syllogistic reconstruction of Euclid I.1 that proceeds backwards, just as Piccolomini does, and ends up with the same four syllogisms as Piccolomini does. We believe this affiliation to be more sound than the one that claims that Clavius

---

[19] Piccolomini (1547: 101b).
[20] Definition 15 reads: "A circle is a plane figure contained by one line such that all the straight lines falling upon it from one point among those lying within the figure are equal to one another" (translation from Heath 1956 [1925]: 183).
[21] Piccolomini (1547: 100b).   [22] Piccolomini (1547: 100b).

was following Herlinus and Dasypodius and thus it is necessary to say something about the latter two.

In 1566 there appeared in Strasbourg a book titled *Analyseis Geometricae sex Librorum Euclidis. Primi et quinti factae a Christiano Herlino: reliquae una cum commentariis et scholiis perbrevibus in eosdem sex libros Geometricos a Conrado Dasypodio*. As the title announced, the book contained an analysis of the first six books of Euclid's *Elements*. The analysis of the first and fifth books was due to Christian Herlinus, while the remaining books were analyzed by Conrad Dasypodius, a student of Herlinus.[23] The analysis in question took the form of a syllogistic reconstruction and this would of course make the book of great relevance to our project. Indeed, the names of Herlinus and Dasypodius recur in the secondary literature on our topic.[24] However, from the logical point of view, the treatment is highly disappointing. While all theorems are presented under a syllogistic garb, the syllogistic reconstruction often does not manage to rise above a superficial restructuring of the steps into two premises (major and minor) and a conclusion.[25] McKirahan (1992) discusses extensively, among other things, the syllogistic version of I.32 given by Herlinus and Dasypodius in the context of a general discussion of whether Aristotle's theory of demonstration can account for Euclidean proofs.[26] McKirahan's discussion is of great systematic interest and its two foci—why can the alleged syllogisms be classified as syllogisms and can the syllogisms truly be combined as syllogistic chains—can be applied to any such reconstruction (and we will do so when analyzing Piccolomini's proof of Euclid I.1 in Section 3.2).

Coming back to Euclid I.1, Herlinus and Dasypodius also use four syllogisms in the reconstruction of that proposition but in the synthetic order (as opposed to the analytic one followed by Piccolomini and Clavius). A parallel reading of their treatment and Clavius' treatment of proposition I.1, indeed, makes it very unlikely that Clavius was following Herlinus and Dasypodius, whereas the similarities between Piccolomini and Clavius are striking (including the use of "resolution"). Could Herlinus and Dasypodius have been familiar with Piccolomini's treatment? We have been unable to discern textual elements that would speak in favor of a direct influence (although this cannot

---

[23] Conrad Dasypodius is well known to scholars interested in the notion of *mathesis universalis*; see Crapulli (1969).

[24] However, the only paper squarely devoted to Herlinus and Dasypodius is Bertato (2014).

[25] A reconstruction of Euclid I.1 worked out by Herlinus and Dasypodius is given in Bertato (2014).

[26] McKirahan (1992: 149–63).

be excluded a priori). That Clavius was following Piccolomini's treatment, rather than Herlinus and Dasypodius, is also supported by the fact that the treatise by Herlinus and Dasypodius is never cited by Clavius in his entire *Opera Mathematica* whereas two works by Piccolomini (but not the *Commentary*) are cited.[27] We need neither engage with a detailed discussion of Clavius' reconstruction nor with that given by Herlinus and Dasypodius, as all the points we make about Piccolomini can easily be applied to them. However, we do want to mention a comment by Clavius concerning the significance of syllogistic reconstructions of Euclid. After his reconstruction of Euclid I.1 Clavius says:

> All the other propositions, not only in Euclid but in all the other mathematicians, can be resolved in the same way. Mathematicians, however, in their demonstrations do not use such resolution, for without it they demonstrate what is required in a shorter and easier way.[28]

Basically, Clavius takes the point of view that while it is in principle significant that all mathematical demonstrations can be handled syllogistically, in practice it would be a waste of time to do so.

Sympathetic to Clavius' position are Pierre Hérigone, Isaac Barrow, and Erhard Weigel. Hérigone treats syllogistically Euclid I.1 in his *Cursus Mathematicus* (1634).[29] He expresses no doubts as to the possibility of syllogizing all mathematical propositions:

> And since every consequence depends immediately on the cited proposition, the proof is obtained from beginning to end through a continuous series of necessary, immediate, and legitimate consequences each contained in a short line, which can all be resolved into syllogisms since in the cited proposition and in that which corresponds to the citation are found all the parts of the syllogism as one can see in the first demonstration of the first book which has been reduced to syllogisms.[30]

---

[27] We thank Professor Eberhard Knobloch for his help on this issue.

[28] Non aliter resolvi poterunt omnes alias propositiones, non solum Euclidis, verum etiam caeterorum Mathematicorum. Negligunt tamen Mathematici resolutionem istam in suis demonstrationibus, eo quod brevius ac facilius sine ea demonstrarent id, quod proponitur [...] (Clavius 1591: 19–20).

[29] Hérigone's discussion is remarked upon in Bertato (2014) and Massa (2010). A recent extensive treatment of Hérigone's *Cursus*, which also devotes attention to the issue of the syllogistic reconstruction of mathematical theorems, is Mellado Romero (2022).

[30] Hérigone (1634: I, Ad Lectorem).

Hérigone proceeded to give his reconstruction of Euclid I.1 in syllogistic terms, which contains four syllogisms and follows the order of the syllogisms given in Herlinus and Dasypodius.[31]

Isaac Barrow offers a syllogistic reconstruction of Euclid I.1 in his lectures on philosophy of mathematics (*Lectiones Mathematicae, lectio* VI, 1664). His reconstruction is, apart from some additional details, very similar to Clavius'. Erhard Weigel, in his *Analysis Aristotelica ex Euclide restituta* (1658), cites Clavius' reconstruction of Euclid I.1 (*Sectio* III, *Membrum* I, *caput* II, §6) and then proposes a 'scholastic' syllogistic reconstruction of Euclid I.32 (*Sectio* III, *Membrum* I, *caput* III, §12).[32] The reconstruction proposed by Weigel goes on for four pages (he begs the reader for patience!) and he concludes by comparing the proof to a four-line proof given in Barrow's edition of Euclid's *Elements*. In the following section he remarks to great rhetorical effect: "The cautious reader will now judge which of the two methods is more expeditious" (§13).[33] Hérigone, Barrow, and Weigel believe that in principle all mathematical demonstrations can be syllogized but, like Clavius, they also believe that in practice there is no usefulness to the exercise. In his 1684 *Meditationes de Cognitione, Veritate & Ideis*, Leibniz also comments, with reference to Herlinus and Dasypodius, on the fact that in practice we do not require such detailed syllogistic reconstructions.[34]

We will not spend more time on Hérigone, Barrow, and Weigel because their presentations do not bring any elements of novelty with respect to Piccolomini's presentation. Just like Piccolomini, they only discuss how the premises of the various alleged syllogisms are to be justified but they do not pay any attention to the structure of the premises and the conclusions nor to the problem of whether the concatenation of the syllogisms results in a syllogism.

At this point we have to evaluate whether Piccolomini's inferences can be called syllogisms. From the outset we reject the line taken by Freguglia in his comments on Clavius' and Piccolomini's syllogistic reconstruction of Euclid I.1:

---

[31] See Massa (2010: 179–82).

[32] There are other attempts at syllogizing (parts of) mathematics in the seventeenth century. We do not discuss them in detail as they do not seem to us to add much to the systematic discussion. See, among such works, Beeckman (1942, vol. II: 170–1) and Sturm (1661).

[33] Weigel (1658: 167): "Judicet nunc prudens Lector utra methodus sit expeditior."

[34] Leibniz (1989: 22) said: "Moreover, a sound demonstration is one that follows the form prescribed by logic. Not that we always need syllogisms ordered in the manner of the schools (in the way that Christian Herlinus and Conrad Dasypodius presented the first six books of Euclid); but at very least the argument must reach its conclusion by virtue of its form."

We cannot agree with those who claim that the logic of the time [Renaissance and seventeenth-century logic] was completely unable or was unsuitable for expressing geometrical demonstrations, in particular the Euclidean ones, since it did not allow for a treatment of relations. First of all, one must observe that, with some stretching, syllogisms can deal with statements that involve mathematical relations. See for instance the equality between segments just as it was treated in the statements present in the syllogisms proposed above by Piccolomini and Clavius. Moreover, those kinds of inferences known as *propositio reduplicativa*[35] and the "oblique syllogism" were also taken into consideration. And although the latter was not properly reducible to Aristotelian syllogism it extended the available deductive structures (exactly concerning relations). Thus, logic, albeit slowly, was developing by considering, in a somewhat less orthodox way, its adherence to the Aristotelian inferential scheme.[36]

Both considerations raise important issues. The first consideration takes its start from the alleged syllogistic reconstruction by Piccolomini and Clavius to conclude that logic was moving toward accepting relational inferences as part of an extended theory of syllogism. But this seems to us to put the cart before the horse! Piccolomini and Clavius had to show that, given syllogistic logic as accepted by the logicians (whatever that amounted to), it was possible to reformulate the Euclidean demonstrations (in particular I.1) as a chain of syllogistic inferences. It is against that standard that the attempt must be evaluated rather than arguing that the attempt is successful because its very implementation has to include forms of reasoning that expand the syllogistic framework. The second consideration also raises important issues and our position is that oblique inferences, while at times involving relations, are not enough to account for the relational inferences appealed to in the syllogistic reconstructions we are discussing. In the next section we will discuss whether Piccolomini's syllogisms can be recast as being constituted of propositions in the appropriate subject-predicate form and what the medium of the syllogisms might be. We will return to oblique inferences when we discuss Vagetius in Chapter 4.

---

[35] This type of proposition is treated, for instance, in Ockham, *Summa Logicae*, 2.16, "De propositionibus reduplicativis in quibus ponitur haec dictio *in quantum*." However, unlike the case of oblique inferences, we are unable to see any connection to mathematical reasoning.

[36] Freguglia (1999: 39); see also Freguglia (1988: 31).

3.2 A Critical Analysis of Piccolomini's Reconstruction

It is now necessary to look at the propositions and the inferences given by Piccolomini and Clavius in more detail. Let us consider, then, Piccolomini's first syllogism:

First Syllogism[37]
Every figure contained within three equal lines is an equilateral triangle;
The figure ABC is contained within three equal lines AB, AC, [BC];
Therefore, the figure ABC is an equilateral triangle.

We must ask in what sense these propositions have the subject-predicate form, what are the extremes, and what is the middle term.

Let M = Figure contained within three equal lines
Let S = being identical to ABC
Let P = being an equilateral triangle.

Here we can reasonably claim to be dealing with a syllogism:

All M is P
all S is M
all S is P.

To achieve this, we need to ignore the difference between "Figure contained within three equal lines" and "contained within three equal lines AB, AC, [BC]" but we could think that if we have established "The figure ABC is contained within the three equal lines AB, AC, [BC]" we can weaken it to "The figure ABC is contained within three equal lines." It stretches things a little but not significantly.

We have given a *Barbara* reconstruction for the first syllogism. Indeed, if we treat ABC as "being identical to ABC" we get three universal premises. An alternative reconstruction can be given by treating propositions with individual terms as particular (i.e., as being of the form "Some A is B"). In that case we get a syllogism of the first figure in the form *Darii*. This is Freguglia's position who reconstructs the syllogism as follows:

---

[37] See Piccolomini (1547: 101a).

all M is P
some S is M
some S is P.

Regardless, the first inference is unproblematically a syllogism.

In the analysis below, we will proceed by treating the syllogisms as *Barbara* but none of our considerations is affected if one prefers to treat them as *Darii*. If the first syllogism seemed to fit well the standard subject-predicate construction of premises and conclusion, we run into serious troubles when trying to recast the second inference in syllogistic terms. Here it is again:

Second Syllogism[38]
Things that are equal to a third are equal to one another;
But AC and BC are equal to the third line AB;
Thus, AC and BC are equal to one another.

How can we parse the premises and the conclusion as categorical statements? The plural quantification "things that are equal to a third" seems to call for a premise of the form All Ps are Qs. But what would be the subject P? Let us try "being equal to a third" for P and "being equal to one another" for Q. But then we must understand the two predicates as being satisfied not by individual objects but by pairs (or pluralities) of lines, such as <AC, AB> (where this way of writing does not necessarily denote a set-theoretical object; it could be a plurality). Following this analysis, we are supposed to find in the second premise an appropriate middle term and an appropriate extreme. If we hope to achieve a reconstruction in *Barbara* we can think of the subject of the second premise as being "<AC and BC>" and for the predicate "being equal to the third line AB." If we look for a middle term for the syllogism, we now have a mismatch. For, the first premise has the predicate "being equal to a third" and the second premise has "being equal to the third line AB." There are two ways in which the mismatch can be eliminated.

The first possibility consists in weakening the first syllogism to a special case of it:

Things that are equal to AB are equal to one another;
But AC and BC are equal to the third line AB;
Thus, AC and BC are equal to one another.

---

[38] See Piccolomini (1547: 101b).

We are apparently able, again with a little massaging of the predicates, to conclude that <AC, BC> are equal to one another (but we argue below that this is only an appearance).

The other possibility is to weaken the second premise to claim only "AC and BC are equal to a third line." The cost of the latter move is that we still need to account for the passage that justifies "AC and BC are equal to a third line" and that justification has to go through establishing "AC and BC are equal to the third line AB."

In either case, we have to alter the premises of the original demonstration (the first premise in the first syllogism and the second premise in the second). This in itself is a major flaw, for we are trying to represent the original proof as a syllogism which takes on board the same premises, not a different proof which contains weakened forms of the premises.

Let us grant, for the sake of argument, that the above moves are successful. We now explain why even with the above reparsing the inferences are not syllogisms. We need to discuss compound terms (such as "Mary and John"). Now "Mary and John love music" can easily be reparsed as Mary loves music and John loves music. But "Mary and John make a good couple" cannot be so reparsed and thus the predicate "make a good couple" would have to apply to the pair, or plurality, <Mary, John>. That is how we have charitably reconstructed the syllogisms above.

But now we run into the problem of establishing the minor of the second syllogism. Piccolomini splits "AC and BC are equal to the third line AB" into two statements: "AC is equal to AB" and "BC is equal to AB." This is what leads to the necessity of two different syllogisms (three and four).[39]

Consider the third syllogism (the same considerations apply to the fourth):

Third Syllogism[40]
All the lines from the center of the circle to the circumference are equal;
But BC and AB connect the center of the circle ACE to the circumference;
Hence, BC and AB are equal.

If we persist in treating "are equal" (or "are equal to one another") as a predicate of pairs, or pluralities, of lines <BC, AB> we lose the syllogistic chain. Assume

---

[39] Incidentally, this also impacts the syllogistic chain because one will need a rule of conjunction introduction to go from "AC = AB" [and] "BC = AB" to "AC&BC are equal to AB." But let us leave that problem for the moment and focus instead on the issue of stating the premises as categorical statements.

[40] See Piccolomini (1547: 101b).

the premise is parsed as "being a pair of lines from the center of the circle to the circumference." We can use the predicate "being equal to one another" to parse the premise as "all pairs of lines from the center of the circle to the circumference are equal to one another." But now the minor cannot be parsed as "everything equal to the pair <BC, AB> connects the center of the circle to the circumference." It is the individual segments that connect the center of the circle to the circumference, not complex objects or pluralities such as <BC, AC>. Thus, in the final analysis, the minor premise of the syllogism requires (and Piccolomini of course knows this) the independent statements "BC connects the center of the circle ACE to the circumference" and "AB connects the center of the circle ACE to the circumference." When we put them together, we do not obtain "<BC, AB> connect(s) the center of the circle ACE to the circumference" but rather "BC and AB connect the center of the circle ACE to the circumference."

Summarizing, the attempt to recast inferences two to four of Piccolomini's proof as syllogistic forces us into two incompatible requests. On the one hand, the universal premises can be recast into subject-predicate form only by appealing to complex terms, such as pairs, or pluralities, of lines. However, the minor premise and the conclusion only work when treating the lines individually.

This can be seen more clearly starting in the inverse order from the one followed by Piccolomini. The first two syllogisms to be established are the third and the fourth. Consider again the third.

Third Syllogism
All the lines from the center of the circle to the circumference are equal.
But BC and AB connect the center of the circle ACE to the circumference.
Therefore, BC and AB are equal.

Premise one is established as an axiom. We could even weaken it to a specific circle:

All the lines from the center of the circle to the circumference ACE are equal.

But in order to be put in subject-predicate form we need to let the predicates range over pairs, or pluralities, of lines. However, to establish the minor premise we need to show, as it is evident by construction, two propositions:

BC connects the center of the circle ACE to the circumference;

and

AB connects the center of the circle ACE to the circumference.

When we put them together (which incidentally is a propositional move and not a syllogistic one) we get

BC connects the center of the circle ACE to the circumference & AB connects the center of the circle ACE to the circumference.

This can for ease be written as

BC and AB connect the center of the circle ACE to the circumference.

But the latter is in no way identical to

<BC, AB> connects the center of the circle ACE to the circumference.

In other words, the surreptitious identification of 'BC and AB' with '<BC, AB>' is responsible for the apparent connection between the three propositions in the alleged syllogisms. And the same problem recurs when one tries to combine the third and fourth syllogism to infer the minor of the second syllogism. In the second syllogism the problem presents itself in the distinction between the need to use compound terms <A, C> and <B, C> to instantiate the major premise and as the subject of the conclusion, whereas the minor needs "AC and BC are equal to AB." But the latter statement is not equivalent to "<AC, BC> is equal to AB."

These considerations are not new,[41] and what we have offered is only one way to look at the problem of dealing with relations in a syllogistic setting. From the contemporary point of view, they amount to nothing else but the realization that the presence of a dyadic relation cannot be straightforwardly accommodated in terms of only monadic predicates. But our line of thought did not depend on knowing that this cannot be done.[42] Rather, it is only meant to show what obstacles are naturally encountered by anyone who will attempt to account for simple relational reasonings with syllogistics. And our point was not to blame Piccolomini or Clavius, or any of the other syllogizers of Euclidean propositions, for not having established the impossibility of such a reduction but rather our concern was to emphasize that they do not show any awareness of what obstacles might stand in the way of such an attempt.

[41] For similar considerations, see McKirahan (1992), chapter 12. McKirahan analyzes Euclid I.32 following Herlinus and Dasypodius.

[42] This would require Church's (1936) theorem to the effect that the set of first order validities with at least one binary relation is not recursive (though monadic first order logic is).

4

Obliquities and Mathematics in the Seventeenth and Eighteenth Centuries

From Jungius to Saccheri

During the fifteenth and sixteenth centuries in Europe, authors like Lorenzo Valla (1407–1457) and Pierre de la Ramée (Petrus Ramus: 1515–1572) considered medieval logic as empty and trifling, and to the scholastic interpretation of logic they opposed the image of a more mundane discipline connected with rhetoric and the art of persuasion. As Jennifer Ashworth remarks:

> The revival of classical scholarship and the turning away from the old style of education led to at least two different approaches among those concerned with logic. Some men concentrated on a return to the pure Aristotle, freed from medieval accretions and misinterpretations; and some turned to a logic which was heavily tinged with rhetoric.[1]

The two approaches—that based on the return to the 'pure Aristotle' and that based on a shift toward rhetoric—persisted during the seventeenth century, together with an attitude witnessed, for instance, by Descartes, Pascal, and Locke, to undermine the importance not only of syllogistic inferences but even of the entire traditional, scholastic logic. Pascal, for example, in his *Art of Persuasion*, wrote:

> The method of not erring is something everyone searches for. Logicians profess to point it out, only mathematicians attain it, and apart from their science and whoever imitates it, there are no true demonstrations.[2]

> I have no doubt, therefore, that these rules, being the true ones, ought to be simple, uncomplicated, and natural, as indeed they are. It is not *barbara* and *baralipton* which make up the argument. The mind must not be

---

[1] Ashworth (1974: 8).   [2] Pascal (1995: 202).

Syllogistic Logic and Mathematical Proof. Paolo Mancosu and Massimo Mugnai, Oxford University Press.
© Paolo Mancosu and Massimo Mugnai 2023. DOI: 10.1093/oso/9780198876922.003.0005

blinkered. Stilted, laboured ways fill it by some foolish presumption with an unnatural loftiness and a vain, absurd pomposity, instead of a solid, vigorous sustinence.[3]

And Locke:

Men can reason well who cannot make a syllogism. If we will observe the actings of our own minds, we shall find that we reason best and clearest, when we only observe the connexion of the proof without reducing our thoughts to any rule of syllogism. And therefore we may take notice, that there are many men that reason exceeding clear and rightly, who know not how to make a syllogism. He that will look into many parts of Asia and America, will find men reason there perhaps as acutely as himself, who yet never heard of a syllogism, nor can reduce any one argument to those forms: and I believe scarce any one makes syllogisms in reasoning within himself.[4]

It would be easy to pile up citations from Bacon, Descartes, and others denigrating the barrenness of syllogisms as aids for finding the truth in any subject matter.[5]

Among the great thinkers of the seventeenth century, only Leibniz speaks in defense of the syllogism:

It must be admitted that the Scholastic syllogistic form is not much employed in the world, and that if anyone tried to use it seriously the result would be prolixity and confusion. And yet – would you believe it? – I hold that the invention of the syllogistic form is one of the finest, and indeed one of the most important, to have been made by the human mind. It is a kind of universal mathematics whose importance is too little known. It can be said to include an art of infallibility, provided that one knows how to use it and gets the chance to do so – which sometimes one does not.[6]

---

[3] Pascal (1995: 204).   [4] Locke (1997: 907).

[5] For the denigration of syllogism in the early modern period, see also Boswell (1991: 126–30), and Petrus (1997: 15–22). For instance, Weigel (1693: *Praefatio ad Lectorem*) writes: "On the contrary, syllogisms are completely unable to find out the truth, a thing which is abundantly confirmed from the experience of many centuries in which this false instrument, which has been given so much credit, has not discovered even one truth. This notwithstanding we do not want logic to be completely rejected or that something be taken away from its praise. It is in fact certain that its use should not be despised when one interprets texts and when one is expounding an already discovered truth."

[6] Leibniz (1981: 478).

Toward the end of the seventeenth century, the interest for the tradition inspired by Pierre de la Ramée decreased, and both the rhetoric and the art of persuasion lost their appeal amongst philosophers and logicians. Even though the logic of traditional scholastic origins never ceased to be taught, the logic textbooks of the time convey a rather impoverished doctrine, mainly centered on a simplified theory of the syllogism.

To this general trend, however, the authors we are now considering constitute an exception. Joachim Jungius recognized that there are some perfectly valid inferences that are non-syllogistic and that are not reducible to a syllogism. Johannes Vagetius, Jungius' disciple, attempted to show that these inferences are necessary to demonstrate mathematical theorems and, in general, logical truths. Leibniz tried to prove the validity of a particular kind of these inferences and Caramuel elaborated a logic based on oblique terms (*Logica obliqua*), that is, terms that imply relations. Saccheri's case is interesting, because he constructs an instance of a valid argument in which relations and nested quantifiers are involved.

These authors can be considered as belonging to a 'movement,' as it were, from traditional logic to mathematics, insofar as they systematically employ traditional logic to prove geometrical theorems (Vagetius) or attempt to improve traditional logic adding to it new types of inferences for performing geometrical and mathematical proofs (Jungius, Caramuel), or assume Euclid's treatment of geometry as a model for developing logic (Saccheri).

To complete our sketchy picture of the status of logic in Europe during the seventeenth century, however, we need to mention another 'movement,' which goes from mathematics toward logic. At the core of this second movement there is the identification of the process of thinking to a *calculus*, which is an absolute novelty with respect to antiquity and the Middle Ages. Petrus Ramus first,[7] and Thomas Hobbes then, with his *Computatio sive logica*, contributed to the development of this idea. In the *Computatio*, indeed, Hobbes writes:

> By ratiocination, I mean computation. Now, to compute, is either to collect the sum of many things that are added together, or to know what remains when one thing is taken out of another. Ratiocination, therefore, is the same with addition and subtraction.[8]

---

[7] According to Nuchelmans (1983: 131), Hobbes "wholeheartedly subscribed to the Ramist view that reasoning is nothing but computation, that is, addition and subtraction." Ramus claimed, indeed, that the Greek word for 'syllogism' denoted properly a calculus or a computation; cf., Nuchelmans (1980: 168–9).

[8] Hobbes (1839, I: 3).

Hobbes' claim may seem quite naïve to us, but it subverted well rooted opinions about the nature of thinking, and paved the way, in the long run, for the establishment of the idea that not only men, but machines as well, can perform the activity of reasoning.

Finally, amongst the factors which contributed to 'drive' mathematics toward logic, a fundamental role was played by François Viète's (1540–1603) discovery of algebra. With the so-called *speciosa generalis*, Viète showed that one can perform calculations involving quantities by operating on letters of the alphabet and attain a high level of generality. The use of letters was perceived at the time as something assimilating algebra to an artificial language, a language particularly fit to express rigorous logical arguments. The idea thus began to circulate that it was possible to express some basic logical operations in algebraic form. Thus, it is no wonder that in 1685 Jacob Bernoulli published a paper entitled *Parallelism between logical and algebraic reasoning*, in which he attempts to compare the way we usually perform logical inferences with an algebraic calculus.[9]

The authors we are considering in this chapter, however, do not belong, as we just remarked, to this second movement, and they are interesting mainly because of their treatment of obliquities and oblique inferences. Thus, let us now turn to Jungius' views on these topics.

Joachim Jungius was born in Lübeck in 1587 and taught mathematics at the Universities of Giessen and Rostock. He studied medicine in Italy at the

---

[9] Cf., Bernoulli and Bernoulli (1685: 213–18). The paper was published under the names of Jacob [Jacques] and Johann [Jean] but it was due to Jacob and it is attributed only to him in Bernoulli (1969). Bernoulli's *Parallelism* has some interesting features which can be summarized as follows:

  (1) Like many other authors at the time, Bernoulli feels obliged to argue in favor of the superiority of algebra over traditional logic.
  (2) He establishes a comparison between some logical operations on the one hand and some algebraic operations on the other.
  (3) He considers concepts as collections of conceptual marks, to which a collection of individual entities, or 'things,' does correspond. Thus, the concept of 'man,' according to this point of view, is a collection of concepts like 'rational,' 'animal,' 'two-footed,' etc., to which a collection of individuals, composed by Peter, Paul, John, etc. is subordinated.
  (4) Bernoulli thinks of the relationship that subsists between a part whatsoever of a concept and the whole concept itself, or idea, as a relation of containment: thus, for instance, the concept corresponding to the word animal is contained or included in the concept corresponding to the word man.
  (5) Bernoulli employs the distinction between comprehension and extension of a concept, derived by Arnauld and Nicole's Logic, and considers the relationship between a concept and its essential 'parts' from the point of view of comprehension.

From the seventeenth to the second half of the nineteenth century in Europe, most of the authors who attempt to reduce logical inferences to a kind of mathematical calculus accept all or some of these features. Thus, Bernoulli's paper, its sketchy form and its lack of depth notwithstanding, may be considered representative of a widespread attitude about the relationship between logic and mathematics which lasted for more than two centuries.

University of Padua and practiced it in Lübeck for four years (1619–23). From 1629 until his death (1657), he was professor of natural sciences at the Akademisches Gymnasium (a secondary school) in Hamburg. A defender of atomism, Jungius considered logic and mathematics as two disciplines indispensable for doing philosophy, and in 1638 published a logic textbook, the *Logica Hamburgensis*, which, even though inspired mainly by the scholastic tradition, contained some very innovative ideas.

Leibniz took Jungius in great consideration and wrote of him:

> And while Joachim Jungius of Lubeck is a man little known even in Germany itself, he was clearly of such judiciousness and such capacity of mind that I know of no other mortal, including even Descartes himself, from whom we could better have expected a great restoration of the sciences, had Jungius been either known or assisted. Moreover, he was already of a mature age when Descartes began to flourish, so it is quite regrettable that they did not know one another.[10]

As Jennifer Ashworth has shown, the most relevant outcome of Jungius' *Logica Hamburgensis* was the explicit acknowledgment that some inferences involving relations cannot be reduced to syllogistic form. Even though these inferences where not completely unknown to medieval logicians (and to Aristotle, as well),[11] Jungius firmly states that they are primitive, have a non-syllogistic nature and are not reducible to a syllogism.

According to a well-established tradition, Jungius defines logic as "an art that directs the operations of our mind to tell the difference of the true from the false."[12] These operations are of three different kinds: the first gives rise to *terms*, the second composes *sentences*, and the third (*tertia mentis operatio*) builds *arguments* (or inferences). Jungius calls 'dianoea'—a calque from the Greek word 'διανοια (= thought, faculty of understanding)'—any type of consequence consisting of an *antecedent* and a *consequent*, regardless of the number of sentences composing the antecedent.[13] Then, he distinguishes the inferences (*dianoeae*) into simple and composite, and defines a *simple*, or *immediate, consequence* (*consequentia simplex*) as that which infers the consequent from the antecedent without interposing any middle term between them.[14] Among the simple consequences he places the *consequence from the right to the oblique* (*a rectis ad obliqua*), of which the following are examples:

---

[10] Leibniz (1989: 7).   [11] Cf., Bochenski (1968 [1951]: 68).   [12] Jungius (1957: 1).
[13] Jungius (1957: 115).   [14] Jungius (1957: 115).

(1) Every circle is a figure;

(2) Therefore, he who describes a circle, describes a figure.[15]

(1) Every walnut is a fruit;

(2) A monkey feeds on a walnut;

(3) Therefore, a monkey feeds on a fruit.[16]

The first of these two inferences is of the same type as the famous one De Morgan employed to show that the traditional syllogism was inadequate to represent certain kinds of valid arguments:

(1) Every man is an animal

(2) Therefore, he who kills a man kills an animal.[17]

Another kind of inference that involves relations and plays an important role in Jungius' logic is the *inversion of relation*, which legitimates the passage from sentences such as 'David is the father of Salomon' to sentences such as 'Salomon is the son of David.'

The point to be emphasized here is that Jungius does not attempt to reduce these inferences to the traditional syllogism, but considers them as separate tools that, together with the syllogism, should improve our capability of reasoning and performing logical proofs. As Alonzo Church remarked in his review of the German edition of the *Logica Hamburgensis*:

These inferences are classed as immediate inferences, along with such traditional inferences as conversion, contraposition and subalternation, as appears in the *tabula dianoearum* (p. 391) and elsewhere. And it is noteworthy that chains of inferences (*dianoeae compositae*) are considered in which these non-traditional immediate inferences constitute one or more of the steps [...] There is brief mention of these inferences in Aristotle (*Topics* 114[a]) but it was Jungius who incorporated them in a systematic way into the body of traditional logic; and apart from Jungius, whose innovation failed of acceptance, they have to be considered non traditional.[18]

---

[15] Jungius (1957: 122).   [16] Jungius (1957: 192).

[17] De Morgan (1847: 114); cf., Sánchez Valencia (1997: 125–6).

[18] Church (1968: 139). In the excerpts that Leibniz took from Jungius' manuscripts, the traditional syllogism is expanded with the inclusion of what Jungius calls 'relative syllogisms,' simple and composite. An example of simple relative syllogism is: "Every figure is greater than any figure inscribed in it (that is, in the figure itself). The circle is a figure. Thus, the circle is greater that any figure inscribed in it (that is, in the same circle)" (Leibniz 1999: 1262).

4.1 Johannes Vagetius (1633–1691)

Johannes Vagetius, Jungius' pupil and author of a long preface to the 1681 edition of the *Logica Hamburgensis*, was well aware of the novelties that Jungius introduced in the body of the traditional logic of Aristotelian origins. Thus, in his preface, he emphasized (perhaps to a fault) the importance of the inversion of relation and of the inference from the right to the oblique.

Animated by a strong desire of showing the superiority of Jungius' logical theory, Vagetius suggests how to overcome the difficulties caused by relational inferences pointed out by Arnauld and Nicole in their *Logic*.[19] Therefore, he attempts to improve Clavius' proof of the first problem in Euclid's *Elements* and devotes several pages to a discussion of Scheibler's treatment of oblique syllogisms proposed in the latter's *Opus logicum*.[20]

Vagetius seems to be well acquainted with Arnauld and Nicole's *La logique ou l'art de penser* that he quotes from the Latin translation of 1666. Introducing chapter 3, devoted to the 'General rules for simple, noncomplex syllogisms,' Vagetius cites *La Logique*:

> This chapter and the following chapters up to chapter 12 are those mentioned in the *Discourse* as containing subtle points necessary for speculating about logic but having little practical use.[21]

Vagetius agrees with the authors of *La logique* that the traditional rules for determining the validity of syllogisms have 'little practical use.' At the same time, however, he observes that Arnauld and Nicole's point of view is too narrow: their claim is true, if we limit ourselves to a consideration of the traditional account of the syllogistic inferences, but if the theory of syllogisms is expanded with other forms of consequences, such as those discovered by Jungius, then it is possible to express in an explicit and correct logical form any kind of proof. Vagetius only worries that, given the generally hostile reaction against various kinds of dogmas typical of his times, traditional syllogistics, if not implemented with the logical rules discovered by Jungius, will soon disappear:

---

[19] Cf., Vagetius (1977: 255). Arnauld and Nicole's *La logique* was first published in 1662.

[20] Christoph Scheibler (1598–1654) taught logic and metaphysics in Giessen for more than twenty years, then theology in Marburg. In 1625 he became superintendent of the *Archigymnasium* in Dortmund. He wrote a huge logic treatise, a part of which discusses the topic of oblique terms and oblique inferences (Scheibler 1654).

[21] Arnauld and Nicole (1981: 182).

Thus, since our time is bothered by the traditionally accepted dogmas, one cannot doubt that many will very soon make the respect hitherto enjoyed by syllogistics disappear unless new forms of consequences are added to the theory of syllogisms: that is, consequences, familiar to the intellect, that have been passed on with suitable principles and that are sufficient, once they are combined with syllogisms, to express in explicit words every argument in demonstrative form.[22]

To show how easily Jungius' logical system can deal with inferences that Arnauld and Nicole consider 'difficult to judge properly,' Vagetius reduces to simple inferences an argument that the two authors of La Logique quote as an example of 'complex argument,' without proposing any analysis of it. For a better understanding of Vagetius' point, it is convenient to quote the entire passage of Arnauld and Nicole:

> But the arguments that are the most difficult to judge properly, and where it is easiest to make mistakes, are those we said earlier can be called complex. This is not just because they contain complex propositions, but because the terms of the conclusion, being complex, are not always joined in their entirety to the middle term in each of the premises, but only to a part of one of these terms. As in this example:
> *The sun is an insensible thing.*
> *The Persians worshipped the sun.*
> *Therefore, the Persians worshipped an insensible thing.*
> Here we see that whereas the conclusion has "worshipped an insensible thing" for its attribute, only a part of it occurs in the major premise, namely, "an insensible thing," and "worshipped" appears in the minor premise.[23]

The difficulty raised by Arnauld and Nicole has to do with the fact that in the first premise of the proposed argument, 'an insensible thing' is a *monadic predicate* attributed to the sun, whereas in the second premise 'worshipped' (or: 'were worshipping,' according to the scholastic usual way of analyzing verbs) is a *relational predicate* that connects two things: the Persians, on the one hand, and the sun, on the other. The conclusion is obtained applying the predicate 'insensible thing' to the second term (the sun) of the relation.

---

[22] Vagetius (1977: 256).   [23] Arnauld and Nicole (1981: 206).

Vagetius proposes two different analyses of this argument:[24]

1.
(1) The sun is an insensible thing,
(2) Therefore, Everyone who worships the sun, worships an insensible thing.
(3) The Persians worship the sun,
(4) Therefore, the Persians worship an insensible thing.

From (1) to (2): inference from the right to the oblique; from (2) to (4) first figure syllogism. To understand how the passage from (2) to (4) works, we need to interpret (2) as 'Every (worshipper of the sun = M) is (a worshipper of an insensible thing = P)' and (3) as 'Every Persian (= S) is (a worshipper of the sun = M)' to conclude: 'Every Persian (= S) is a worshipper of an insensible thing = P).'

2.
(1′) The Persians worship the Sun,
(2′) Therefore, the Sun is worshipped by the Persians.
(3′) But the Sun is an insensible thing,
(4′) Therefore some insensible thing is worshipped by the Persians.
(5′) Therefore the Persians worship some insensible thing.

From (1′) to (2′): inversion of relation; from (2′) to (4′) third figure syllogism (with an analogous maneuver as before); from (4′) to (5′) inversion of relation.

It is Vagetius himself who explicitly justifies the logical steps of these two arguments as we have reported them.[25]

Clearly, Vagetius is eager to show the importance of both the inference rules '(re-)discovered' by Jungius, and this explains why he proposes *two* different analyses of the same argument: the first employing the inference from the right to the oblique and the second appealing to the inversion of relation. At any rate, it is also quite clear that what he calls 'syllogisms' are not syllogisms at all, if with 'syllogism' one means the traditional Aristotelian syllogism.

After having discussed the argument proposed by Arnauld and Nicole, Vagetius suggests how to mend Clavius' proof of the first problem of Euclid's *Elements*. To this aim, he first faithfully quotes Clavius' proof and

---

[24] Vagetius (1977: 257).
[25] We have only added the numbers on the left and disposed the justifications under the arguments. Instead of numbers, Vagetius uses braces on the right to gather in a group the single steps.

then develops his own analysis, which, according to him, should improve that of Clavius. Vagetius first quotes Euclid's problem: "On a given straight line of finite length construct an equilateral triangle." Then he reminds the reader of what, according to the Euclidean tradition, a *problem* properly is:

> Every problem implies an operation and tells us how to do, to construct, to find something, and for this reason has two parts: a proposition stating what has to be done and a construction, by means of which is explained how the construction has to be done. And because every action or activity concerns singular things, this is the reason why problems, that for the most part are proposed in a universal way, are usually restricted by means of ecthesis to singulars.[26]

Concerning the notion of *ekthesis*, Vagetius (as Risse remarks in his commentary to Vagetius) depends on Jungius: "an ecthetic proof is such that its propositions are singular, but so formed that our understanding grasps them as universal."[27] Thus, once considered from the point of view of *ekthesis*, Euclid's problem becomes: "On a given finite straight line AB construct an equilateral triangle." And transformed into a theorem:

> If, on the finite straight-line AB, assuming as centers the two points A and B we describe two circumferences, or arcs, intersecting themselves on point C and then we trace the straight lines AC, BC, then the triangle ACD so constituted is equilateral.[28]

Thus, the proposition to be demonstrated is that '*the triangle ABC is a triangle having the three sides equal among themselves.*' Vagetius' demonstration rests on a Definition and an Axiom:

[Def.] "Every triangle with three sides equal among themselves is equilateral";
[Axiom]: "Things that are equal to the same thing are equal among themselves."

Vagetius' proof proceeds as follows.[29] First a syllogistic proof of the theorem is stated:

---

[26] Vagetius (1977: 257).   [27] Vagetius (1977: 257); see also Jungius (1957: 224).
[28] Vagetius (1977: 258).   [29] Vagetius (1977: 260).

(1) Every triangle with three sides equal among themselves is an equilateral triangle.
(2) The triangle ABC is a triangle with three sides equal among themselves.
(3) Therefore: The triangle ABC is an equilateral triangle.

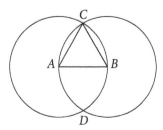

Then, Vagetius applies himself to prove premise 2 of the syllogism, dividing the task in several steps. Employing the Axiom, he constructs the following first figure syllogism:

(4) Things that are equal to the same thing are equal among themselves;
(5) Things equal to the side AB are equal to the same thing;
(6) Therefore, things that are equal to the side AB are equal among themselves.

Proposition 6 (the conclusion of the previous syllogism) becomes the premise of another syllogism:

(7) Things that are equal to the side AB are equal among themselves;
(8) Now, the straight lines AB, BC, AC are equal to the side AB;
(9) Therefore, the straight lines AB, BC, AC are equal among themselves.

Finally, the conclusion (9) becomes the first premise of this syllogism:

(10) The straight lines AB, BC, AC are equal among themselves;
(11) But the three sides of the triangle ABC (by construction) are the straight lines AB, BC, AC;
(12) Therefore, the three sides of the triangle ABC are equal among themselves.

The sentence at step 12 is grammatically equivalent to the sentence expressing the second (minor) premise of the first syllogism (step 2 above): "The triangle ABC is a triangle with three sides equal among themselves." Thus, the proof of the second premise of the main syllogism is concluded.

At this point, Vagetius remarks that the "minor premise of the intermediate syllogism at steps 7–9 is composed by three members: 1) AB is equal to AB; 2) BC is equal to AB; 3) AC is equal to AB."[30] Consequently, he attempts to prove each of these 'members':

The minor premise of the intermediate syllogism 7–9 consists of three members: (1) AB is equal to AB; (2) BC is equal to AB; (3) AC is equal to AB.

 I. [First] member.
    (13) The same is equal to itself,
    (14) Hence AB is equal to the side AB.
 II. Def. 15. All the straight lines drawn from the center to the periphery are equal among themselves,
    (16) All the straight lines drawn from the center A to the periphery CB are straight lines drawn from the center to the periphery,
    (17) Therefore, all the straight lines drawn from the center A to the periphery CB are equal among themselves.
    (18) Now, the straight lines AB, AC are straight lines drawn from the center to the periphery BC,
    (19) Therefore, the straight lines AB, AC are equal,
    (20) Therefore, the straight line BC is equal to the straight line AB.
III. All the straight lines drawn from the center to the periphery, etc.
    (21) Therefore, all the lines from the center B to the periphery CA, etc., as in the case of the second member.

Vagetius considers the second premise (step 5) of the syllogism 4–6 as the result of the application of the *consequence from the right to the oblique* to the quite odd sentence "The side AB is something identical (*Latus AB est aliquid identicum*)." Clearly, Vagetius attributes the same logical structure to the following two inferences:

(a) (typical example of the *consequence from the right to the oblique*)
Painting is an art;
He who learns painting learns an art.

(b)
The side AB is something identical;
Therefore, what is equal to the side AB is equal to something identical.

---

[30] Vagetius (1977: 261).

The predicate "something identical" applied to the side AB is a predicate artificially concocted by Vagetius with the aim of uniformizing the inference (b) to a case of a consequence from the right to the oblique, like (a).

Vagetius' proof has the same structure as Clavius' proof. The main inference has the general form:

(A) Every triangle whose sides are equal is equilateral;
(B) The triangle ABC has three equal sides;
(C) Therefore the triangle ABC is equilateral.

Both Clavius and Vagetius concentrate their efforts to prove (B), the minor premise of this argument, and it is at this point of the proof that Vagetius attempts to show the superiority of his method. Clavius, indeed, according to Vagetius' reading, employs this argument to prove (B):

(α) Things that are equal to the same are themselves equal; (Axiom).
(β) The two sides AB, BC are equal to the same (i.e., the side AB);
(γ) Therefore, the two sides AB, BC are themselves equal.
(δ) Because of this (and consequently) the three sides AB, BC, AC are all equal.

Vagetius criticizes Clavius' demonstration because it is too contracted and with many gaps. This is the reason why he proposes a new demonstration in which all gaps are allegedly filled. Concluding his amended proof of the first problem of Euclid's *Elements*, Vagetius remarks:

> This resolution – free from gaps and without any substitution of syllables or letters in place of others, unless carried out in the form of a certain conse- quence recommended by this logic, – deduces the conclusion from its principles by means of a continuous series of coherent reasonings. Consequently, no errors can be feared to appear in it, thereby making its comprehension absolutely certain.[31]

Vagetius' ideal of a mathematical proof is that of a continuous chain of 'coherent reasonings,' which, without gaps, leads from the axioms or principles to a conclusion. It is interesting to note the very general character of this remark: the syllogism is not assumed here as the paradigmatic inference of the scientific demonstration; in fact no mention is made of the syllogism altogether.

---

[31] Vagetius (1977: 261).

In his Preface to the *Logica Hamburgensis*, Vagetius devotes several pages to criticizing Scheibler's treatment of oblique syllogisms. Scheibler, for instance, claims that any syllogism having a premise with an obliquity must have the conclusion with an oblique case too. If either the major or the minor term (one of the two 'extremes') is oblique in one of the premises, it must be oblique in the conclusion. Vagetius shows with counterexamples that this claim is wrong and that the same holds for other rules proposed by Scheibler. As Leibniz—who read the Preface immediately after its publication—remarked: "Vagetius [. . .] tests Scheibler's rules and shows that they are not too accurate."[32]

An important aspect of Vagetius' investigation into Scheibler's treatment of obliquities is his clear acknowledgment that, to determine the right behavior of oblique inferences it is necessary to consider the possibility of quantifying the predicate:

> Incidentally, it is worth remarking here that we need to determine the quantity of the predicate as well, or of the syncategorematic parts, if we must specify the rules to be employed when we want to judge about the obliques without fear of making mistakes.[33]

In his account of Vagetius' Preface, Leibniz did not miss this point and finds "useful to determine the quantity of the predicate."[34]

4.2 Gottfried Wilhelm Leibniz (1646–1714)

Vagetius was a correspondent of Leibniz and they discussed together, among other things, the inferences from the right to the oblique.[35] On April 4, 1687, Vagetius sent a letter to Leibniz, where he attempted, in a quite convoluted way and employing the *inversion of relation*, to prove the inference from the right to the oblique. Leibniz answered stating that he agreed with Vagetius'

---

[32] Leibniz (1999, 4B: 1120). After having examined Scheibler's treatment of obliquities, Vagetius comes back to the *Logique* of Arnauld and Nicole, still continuing his personal crusade in favor of the new inferences proposed by Jungius. This part of the Preface, however, is quite wordy and unexciting and we don't consider it of particular interest in connection with the aim of the present chapter.

[33] Vagetius (1977: 277).   [34] Leibniz (1999, 4B: 1120).

[35] Leibniz repeatedly asked Vagetius for permission to peruse the papers left unpublished by Jungius, after his death. A great part of Jungius' unpublished papers, however, were destroyed at the end of the seventeenth century, when a fire broke out in Vagetius' house, where they were preserved. Fortunately, Leibniz made excerpts from them before their destruction and so we are now able to read about one hundred pages of Jungius' logical work through the transcriptions made by Leibniz. Cf., Antognazza (2009: 209). For Leibniz's transcriptions, cf., Leibniz (1999, 4B: 1211–306).

proof and said that we must always try to reduce complex things to simpler ones, and that, in particular, when we have not reached the simplest elements, different possible analyses are at our disposal (which was in essence an elegant way of undermining Vagetius' analysis).[36] Thus, Leibniz himself proposed a reduction by means of a proof of the inference from the right to the oblique, based essentially on the principle of substitutivity.[37]

Leibniz's proof rests on three *suppositions*, two of a purely logical kind and the third merely grammatical.

First supposition. Given a true universal affirmative sentence of the general form 'Every S is P,' the predicate 'P' can be substituted at the place of the subject 'S' in every true sentence where 'S' plays the role of the predicate. Thus, for instance, given these two sentences:

(1) Every S is P (every human being is mortal);
(2) a is S (Socrates is a human being),
according to the principle, we may conclude:
(3) a is P (Socrates is mortal).[38]

Leibniz considers even the converse of this rule, according to which "the predication follows from the substitution," and thus distinguishes two cases:

> For example, (first part) since painting is an art, if we have 'a thing which is painting' we shall be able to substitute 'a thing which is an art'. Conversely, (second part) if it should be shown that for 'he who learns painting' it is always possible to substitute in the way described, without loss of truth, 'he who learns an art', then the proposition 'He who learns painting, learns an art' will be true.[39]

Second supposition. The grammatical 'supposition' states the (semantic) equivalence of expressions like:

(a) he who learns a thing which is X (painting; an art);[40]
(b) he who learns X (painting; an art).[41]

---

[36] Leibniz (1768, 1: 37).

[37] Leibniz's entire letter has been translated into English by George H. R. Parkinson in Leibniz (1966: 88–9).

[38] Cf., Leibniz (1966: 88): "To be a predicate in a universal affirmative proposition is the same as to be capable of being substituted without loss of truth for the subject in every other affirmative proposition where that subject plays the part of predicate; from a predication, therefore, this substitution follows, and conversely the predication follows from the substitution."

[39] Leibniz (1966: 88).

[40] Cf., Leibniz (1768, 1: 37): "*qui discit rem, quae est graphice* [...] *qui discit rem quae est ars* [...]."

[41] Leibniz (1768, 1: 37): "*qui discit graphicen* [...] *qui discit artem.*"

(a) and (b) being semantically equivalent, can be substituted for one another *salva veritate*.[42] In this case, too, Leibniz distinguishes two parts:

> A general oblique case taken with a particular direct case is equivalent to a particular oblique case, and therefore they can be substituted mutually for one another. For example, (first part) 'he who learns a thing which is painting' can be substituted for the term 'he who learns painting'. Conversely, (second part) 'he who learns an art' can be substituted for the term 'he who learns a thing which is an art'.[43]

(a) is an instance of a *general oblique case, taken with a particular direct case,* because the term 'thing (*res*),' which is in an oblique case (accusative), has a very general reference and is followed by a relative ('which' [*quae*]) in nominative (direct case). (b), instead, is *a particular oblique case,* because the obliquity concerns a specific term ('painting' or 'art'), not a generic one, as 'thing.'

Third supposition. The third 'supposition' is the principle of transitivity applied to substitutivity:

If B can be substituted for A, and C for B, and D for C, then D can also be substituted for A.[44]

Once the three suppositions have been accepted, Leibniz proposes the following proof:

Theorem
<The inference> from the direct to the oblique is valid.
Article 1. Grant, in the direct case, 'Painting is an art.'
Article 2. I assert that there follows from this, in the oblique case, 'He who learns painting, learns an art.'

Proof
Article 3. Let there be a term 'he who learns painting.'
Article 4. 'He who learns a thing which is painting' can always be substituted for this (by the first part of the second supposition).
Article 5. Again, there can always be substituted for this, in the manner stated above, 'he who learns a thing which is an art' (by the first part of the first supposition). For painting (article 1) is an art.

---

[42] Leibniz (1768, 1: 37).   [43] Leibniz (1966: 88).   [44] Leibniz (1966: 88).

Article 6. But for this, again, 'he who learns an art' can always be substituted (by the second part of supposition 2).

Article 7. By the third supposition, therefore, arguing from the first to the last (namely, from article 3 to article 6, through 4 and 5), for the term 'he who learns painting' there can always be substituted, in the manner stated above, 'he who learns an art.'

Article 8. From this, finally, there is inferred (by the second part of supposition 1) 'He who learns painting, learns an art.'

Q.E.D., as was proposed in article 2.[45]

Leibniz attributed directly to Jungius the discovery of the non-syllogistic inference from the right to the oblique and of the inversion of relation, and believed that, even though they were not reducible to syllogisms, they could be proved by appealing to more general principles upon which even the traditional syllogisms depend:

> It should also be realized that there are *valid non-syllogistic inferences* which cannot be rigorously demonstrated in any syllogism unless the terms are changed a little, and this altering of the terms is the non-syllogistic inference. There are several of these, including arguments from the right to the oblique – e.g. 'If Jesus Christ is God, then the mother of Jesus Christ is the mother of God'. And again, the argument-form which some good logicians have called inversion of relation, as illustrated by the inference: 'If David is the father of Solomon, then certainly Solomon is the son of David'. These inferences are nevertheless demonstrable through truths on which ordinary syllogisms themselves depend.[46]

Probably, when Leibniz says that these inferences are demonstrable, he is thinking of some proof similar to that included in the letter to Vagetius which we have just considered.

Thus, to summarize, Leibniz has a twofold attitude toward the inference from the right to the oblique and the so called 'inversion of relation': on the one hand, he believes that they are *non-syllogistic* and not reducible to the syllogism; on the other hand, he thinks that they can be proved on the basis of very general principles on which the syllogism itself rests. In Leibniz's works, besides the letter to Vagetius, there are only few scattered references concerning the nature of such a proof. In a text composed in 1686, for instance,

---

[45] Leibniz (1966: 88–9).   [46] Leibniz (1981: 479).

Leibniz hints to a 'resolution' of oblique cases "into several propositions" as a necessary condition for the proof:

> If you do not resolve oblique cases into several propositions, you will never find any way out, unless, with Jungius, you will be forced to imagine some new modes of reasoning.[47]

In another paper belonging to the same period, he claims that to perform the proof we must bring back the 'Jungian' inferences to the *Characteristic grammar*, that is, they must undergo some kind of 'grammatical analysis':

> Those consequences that Jungius noticed and cannot be proven by any syllogism or other logical device must be brought back to Characteristic grammar.[48]

This agrees with what Leibniz says in the passage quoted above, according to which to demonstrate these inferences we need to change a little the terms and that it is exactly this "altering of the terms" that makes the inference non-syllogistic. Concerning the details of these grammatical transformations, however, we are left in the dark.

At any rate, Leibniz elaborated a very broad and 'liberal' idea of *logic* and of *logical form*, as witnessed by the following passage, which we have already partially quoted, from the *New Essays*:

> It must be admitted that the scholastic syllogistic form is not much employed in the world, and that if anyone tried to use it seriously the result would be prolixity and confusion. And yet – would you believe it? – I hold that the invention of the syllogistic form is one of the finest, and indeed one of the most important, to have been made by the human mind. It is a kind of universal mathematics whose importance is too little known. It can be said to include an *art of infallibility*, provided that one knows how to use it and gets the chance to do so – which sometimes one does not. But it must be grasped that by 'formal arguments' I mean not only the scholastic manner of arguing which they use in the colleges, but also any reasoning in which the conclusion is reached by virtue of the form, with no need for anything to be added.

---

[47] Leibniz (1999, 4A: 115): "*Nam nisi obliquos casus resolvas in plures propositiones numquam exibis quin cum Jungio novos ratiocinandi modos fingere cogaris.*"
[48] Leibniz (1999, 4A: 120).

So: a sorites, some other sequence of syllogisms in which repetition is avoided, even a well drawn-up statement of accounts, an algebraic calcula-tion, an infinitesimal analysis – I shall count all of these as formal arguments, more or less, because in each of them the form of reasoning has been demonstrated in advance so that one is sure of not going wrong with it. Most of Euclid's demonstrations, too, are close to being formal arguments.[49]

Before concluding this section, we think it will be rewarding to insist a little more on Leibniz's theory of relations. To this aim it is necessary to separate the ontological issue (what is the proper nature of relations and relational proper-ties) from the logical one (how can we handle relational sentences, i.e., sentences including relations when we are making logical inferences). Let us first begin with the ontological interpretation of relations.

According to the scholastic tradition, on which Leibniz heavily relies, relations understood as properties connecting two or more individuals do not belong to the 'furniture of the world.' Thus, if Solomon and David are two existing individuals and Solomon is the son of David, what exists in the 'external world' are two individuals called respectively 'Solomon' and 'David' with their internal properties (being human, rational, capable of singing, etc.), not the properties corresponding to the words 'being a father' or 'being a son' conceived as something common to both these individuals. This point of view is synthetically expressed by the old adage that the same accident cannot belong to two different subjects (or individual substances).[50] Leibniz expresses the same concept in a picturesque way, saying that there are no accidents with a leg in a subject and a leg in another.[51]

The prevailing scholastic (and late-scholastic) doctrine about the nature of (binary) relations was that any relation splits up into two relational properties each inhering in a subject. If Solomon is the son of David, David has the property of being a father and David that of being a son. This emerges with great clarity from a text where Leibniz reports Jungius' opinion concerning the nature of relations:

[Jungius] says that there is not a *Relation* between the related things, but that there are so many relations as there are related things, i.e. a relation is that in

---

[49] Leibniz (1981: 478–9).

[50] Cf., Thomas (1980: I, 206 b); Burleigh (1955: 33); Toletus (1579: 83 v). On medieval theories of relations, see Henninger (1989) and Mugnai (2016).

[51] Leibniz (1969: 704); Mugnai (1992: 30).

virtue of which A is said to be similar to B, and another relation is that according to which B is said to be similar to A.[52]

Worth noting is the fact that the refusal of attributing reality to the relation understood 'as a bridge' connecting the related things was shared by most of the philosophers of the Western tradition, independently of their ontological position. On this point, the 'realist' Walter Burley (also spelled Burleigh) was on the same side as the nominalist Ockham or Buridan. The dispute about the reality of relations, was properly a dispute concerning the reality of what we may now call *relational properties*, that is properties like 'being a father of...,' 'being older than...,' which belong to a given subject and are conceived as properties inhering in it. In particular, at Leibniz's times, the dispute concerned the relationship between the *relational properties* and the *absolute (non-relational) properties* on which they are supposed to be grounded. If Socrates and Plato are similar, to pursue Jungius' example, they are similar because of some property they share; and this property is exactly what, according to the ontology of Aristotelian tradition, was called 'foundation' of the similarity relation. Now, suppose that both Socrates and Plato are philosophers: then the dispute between nominalists and realists concerned the question whether the *relational property* of *being similar* inhering in Socrates was something *really distinct* from the property of being a philosopher or not.

Leibniz, on his part, believed not only that the relational property was not really distinct from its foundation, but he was even persuaded that relations conceived as something *in between* the related items were *merely mental things*. Interestingly enough, there are texts where he openly claims that it is precisely because relations are understood as properties inhering in a plurality of things, that they must have a mental nature. In other words: it is their *polyadicity* that reveals the intrinsic mental nature of relations.

Leibniz's position strongly resembles that of Peter Auriol (c.1280–1322), a scholastic thinker active in the early part of the fourteenth century:

[...] a relation is only in apprehension, having no being in things, because that which exists as one and simple and connects [attingit] two really distinct things seems to be only the work of the intellect, otherwise the same simple and individual thing will be in several things separated from each other. But it is clear that a relation connects two distinct things, one as a foundation and

---

[52] Leibniz (1999, 4B: 1067).

the other as term. Then, being something indivisible and simple, it cannot be in the extra mental reality, but only in the consideration of the intellect.[53]

The important novelty of Auriol's and Leibniz's position, as regards the philosophical tradition of scholastic origin, is the acknowledgment that there are relations connecting a plurality of things, even though their nature is restricted to the realm of mental beings. This restriction was necessary to save the underlying Aristotelian ontology of a world composed only of individual substances with their internal properties and without accidents common to two or more substances.

Once established that relations are not *real*, we may think that Leibniz feels free to employ them without problems, considering, as Bertrand Russell remarked, that as a mathematician, he was perfectly aware of their importance. In his essays on geometry and *analysis situs*, Leibniz systematically employs relations and formal properties of relations to develop a kind of 'qualitative' geometry. In a text on geometry, for example, Leibniz proposes to represent as follows the fact that the points A and B on the Euclidean plane maintain a unique relation with the point L: *A.B.L.un.* As he explains, this means that if *A.B.C.un.* holds, then C = L holds too.[54] In other works, for the most part unpublished, Leibniz employs various kinds of symbolism to represent geometrical relations and at least in a paper on *Mathesis universalis* he proposes to introduce a symbol for representing "a relation in general."[55] Moreover, at the end of his life, in a marginal note to a book on metaphysics, Leibniz has no problems at all to admit that there are 'relations of relations,' as well: "On the contrary: two similarities can be similar."[56]

It is when working in the field of logic, however, that Leibniz has a more conservative attitude toward relations and seems to be forced to take the traditional ontology into account. Thus, for instance, given a sentence composed of two subjects and a relation connecting them, Leibniz proposes to analyze it by means of a conditional joining two other sentences in subject-predicate form. So, given the sentences:

(1) 'Paris loves Helen'
(2) 'Titius murders Caius'

---

[53] Auriol (1596–1605). In I *Sent.*, d. 30; cf., Mugnai (2016: 537–8).
[54] Here Leibniz seems to have a clear insight into the notion of *function*; cf., Schneider (1988: 181ff.) and Mugnai (1992: 90–2).
[55] Leibniz (1971, 7: 57).   [56] Mugnai (1992: 163, note 87).

Leibniz reduces them, respectively, to:

(3) 'Paris is a lover and exactly for that (*et eo ipso*) Helen is loved'[57]
(4) 'Insofar as Titius is murdering, therefore Cajus is murdered.'[58]

Clearly, he wants to avoid considering as primitive a sentence in which two subjects are joined together by a relation.

To conclude, we may say that Leibniz has a very broad notion of what a demonstration is, but that, at the same time, he has a quite traditional (and therefore narrow) idea of the logical devices needed to perform it. Even though he agrees that 'Jungian' inferences cannot be reduced to syllogisms, he doesn't accept as primitive the inversion of relation and other overtly relational sentences. Thus, in several papers he attempts to reduce the relational sentences to a more fundamental logical structure based on sentences in subject-predicate form linked by appropriate syncategorematic terms.[59]

4.3 Juan Caramuel Lobkowitz (1606–1682)

At the very beginning of his *Logica obliqua*, published in 1654 as part of his *Theologia Rationalis*, Caramuel mentions the various kinds of oblique propositions that he will discuss in the treatise. These are propositions of mathematics and geometry, propositions containing comparatives and superlatives, and expressions referring to places and times. Caramuel proposes a quite original analysis of the structure of an elementary proposition with obliquity. Given, for instance, the sentence 'Ferdinand is a prince appreciated by the people,' he considers 'Ferdinand' and 'people' as the two terms of the relation and 'is a prince appreciated by' as the copula; analogously, in the case of 'Peter could come quickly to Rome,' 'Peter' and 'Rome' are the terms of the relation and 'could come quickly [to]' the copula, which, according to him, 'is composed by three words.'[60]

Caramuel recognizes three different types of obliquity:

(1) when the first term playing the role of subject implies an oblique case, as in 'a soul which committed sin will die ['sin' is accusative]';

---

[57] Leibniz (1999, 4A: 114–15).   [58] Leibniz (1999, 4A: 651).   [59] Leibniz (1999, 4A: 651).
[60] Caramuel (1654: 408).

(2) when the second term, playing the role of a predicate, is in an oblique case, as in 'the soul aspires to glory' [*ad gloriam*: accusative];

(3) when both terms (subject and predicate) involve an obliquity, as in 'the soul which committed a sin will be sentenced to eternal damnation.'[61]

In the *Logica obliqua*, Caramuel concentrates mainly on the second type of oblique sentences, 'with some treatment' of sentences of type one.[62]

Discussing oblique terms, Caramuel identifies a peculiar class of relations, which have the property of being *symmetric*, that is, such that, in contemporary notation, if R(x, y) holds, then R(y, x) holds as well. Appealing to a traditional, scholastic classification of relations, Caramuel distinguishes between relations that are based on the same property of the related items (*relationes aequiparantiae*) and relations that attribute different properties to them (*relationes disquiparantiae*).[63] To the first class belong relations like 'being similar,' 'being a brother of,' 'being married to...,' 'being near to...,' and so on; to the second kind of relations those implied in the verbs 'to give,' 'to beget,' 'to envy,' and so on. If John is married to Mary, it follows that Mary is married to John; if John is a brother of Philip, then Philip is a brother of John, whereas if John gives something to Mary, from this it does not follow that Mary gives something to John.

In contemporary notation, for 'R' a relation and x, y, individuals whatsoever, we have:

(a) $(x)(y)(R(x, y) \rightarrow R(y, x))$ — 'R' is a *symmetric* relation;

(b) $\neg (x)(y) (R(x, y) \rightarrow R(y, x))$ — 'R' is *asymmetric*.

Thus, as Caramuel remarks, the following inferences are all valid:

(1) Caius is the brother of Peter; therefore, Peter is the brother of Caius;

(2) Mary is married to Anthony; therefore, Anthony is married to Mary;

(3) Gneius is near Claudius; therefore, Claudius is near Gneius.[64]

---

[61] Caramuel (1654: 407).   [62] Cf., Dvorák (2008: 655ff.).   [63] Caramuel (1654: 409).
[64] Caramuel (1654: 409).

Whereas inferences like the following are all invalid:

(4) A disease killed Peter; therefore Peter killed a disease;

(5) John is looking at a blind man; therefore a blind man is looking at John

(6) A river floods the town; therefore the town floods a river.[65]

Indeed, in (1), (2), (3), the relations involved are *symmetric relations (relationes aequiparantiae)*, whereas in (4), (5), (6) they are *asymmetric relations (relationes disquiparantiae)*.

Thus, Caramuel states the following rule:

> *If a relation is symmetric [aequiparans], then its conversion is performed according to the simple [usual] way; if it is asymmetrical [discrepans], then it is performed according to the same way as well, but a single sign of negation [unum non] must be added to the converted proposition, which, affecting the copula, transforms the proposition itself into a negative one.*[66]

Thus, according to this rule, the proposition 'Andrew is an apostle of God' can be converted into 'God is not an apostle of Andrew,' because 'being an apostle,' like 'being a servant,' is an *asymmetrical* relation.[67] In other cases, however, the conversion is possible by means of a change from the active to the passive. A proposition like 'Milo kills Claudius,' for instance, can be converted into 'Claudius is killed by Milo.'[68] However, there seems to be an unwarranted move in Caramuel's treatment of asymmetrical relations. While it is correct to say that when the relation is asymmetrical one cannot infer from Rxy that Ryx, it is also not correct, contrary to what Caramuel seems to believe, to go from Rxy to not: Ryx. From John gives something to Mary it does not follow that Mary does not give something to John. Only if R is never symmetric, would Caramuel's inferences go through; but then 'to give,' 'to envy,' and so on would not be examples of such relations.

As Caramuel emphasizes, the distinction between the two kinds of relations was well known to the scholastic and late-scholastic thinkers, tracing back to Aristotle's commentaries on the *Categories*.[69] To the best of our knowledge,

---

[65] Caramuel (1654: 409).   [66] Caramuel (1654: 410).   [67] Caramuel (1654: 410).

[68] Caramuel (1654: 410).

[69] The distinction was known with different Latin names: *relationes aequiparantiae* versus *relationes disquiparantiae*; *relativa univocationis* versus *relativa aequivocationis*; *relata aequalis comparationis* versus *relata inaequalis comparationis*.

however, Caramuel was the first to link this distinction with a logical distinction between a proposition and its converse.

Another important novelty of Caramuel's treatise is the attempt to find criteria for the equipollence of oblique propositions containing quantified expressions and negation. Thus, for example, Caramuel claims that 'Every saint loves every neighbor' is equipollent with 'No saint doesn't love every neighbor'; and 'Every man has some bad habit' is equipollent with 'There is no (some) man, who does not have some bad habit.'[70] To investigate logical equivalences of this kind, he proposes a peculiar symbolism aimed at distinguishing the various parts of a proposition and, in particular, to indicate the scope of negation.[71]

When discussing inferences, Caramuel too, just as Ockham and Buridan before him, selects only inferences that hold in virtue of the *dictum de omni et nullo*, that is, of the monotonicity principles governing the usual quantifiers of categorical propositions. This is an example:

We are looking carefully at some dogs;
Every dog is an animal capable of barking;
Therefore, we are looking at some animals capable of barking.[72]

Caramuel's views concerning the copula and the structure of elementary sentences resemble those held later by Augustus De Morgan. The differences between the two authors, however, are more evident than the similarities. Caramuel, for instance, attempts to integrate obliquities into the syllogism and constructs a general type of genuinely oblique syllogism that, he claims, has a 'Platonic form' (the name is due to the desire of distinguishing the new oblique syllogisms from the old 'Aristotelian' ones). According to Caramuel, an example of syllogism in 'Platonic form' is:

(1) Three elements are less than four [dative] elements;
(2) Four elements are less than six [dative] elements;
(3) Therefore, three elements are less than six elements.[73]

Conditions for the validity of this syllogism are: the middle term must express some obliquity in the major premise and must be in the nominative case in the minor premise; the subject of the conclusion must be in the

---

[70] Caramuel (1654: 411).   [71] Caramuel (1654: 411).
[72] Caramuel (1654: 430).   [73] Caramuel (1654: 432).

nominative case in the premise in which it occurs and in the conclusion; the predicate of the conclusion must be in an oblique case in the conclusion itself and in the minor premise.[74]

All the examples of 'Platonic syllogism' proposed by Caramuel involve transitive relations and clearly Caramuel thinks of this kind of syllogism as an inference that preserves transitivity. These examples, for the most part, are instances of the general schematic form: 'If Rxy and Ryz, then Rxz,' where 'R,' the relation involved, is not *identity*. To understand this point, we need to consider that in the *Logica obliqua* Caramuel claims full originality for his treatment of oblique sentences:

> I will transmit now a new logic, i.e. an oblique logic about which the ancient dialecticians said nothing or very little. I will transmit, I say, a logic completely different from the ancient one and resting on opposite grounds.[75]

According to him, the main difference with respect to the old logic is that the new one denies the principle on which the traditional theory of syllogistic inferences was supposed to rest: 'if two things are equal to a third thing, then they are equal to each other.' Caramuel observes that this principle was considered as an article of faith ('the expression of an oracle') by the followers of Aristotle and, in evident opposition to them, he does not hesitate to state that it is false and misleading. Oddly enough, however, he does not say what principle (or principles) he substitutes for it and he merely states that in the new logic he will employ terms 'really different' (i.e., terms that do not permit the application of the principle of transitivity of identity). For this reason, to characterize the new logic, he coins the expression 'discrete logic [*logica discreta*],' aiming to suggest that the inferences in this kind of logic are performed without applying transitivity to the identity relation.[76]

Caramuel, however, does not recognize explicitly the property of transitivity as such and instead keeps on displaying a series of concrete instances of oblique inferences that reproduce the same pattern. The first figure oblique syllogism is easily obtained by the 'Platonic syllogism,' inverting the order of

---

[74] Caramuel (1654: 432). Thus stated, however, these conditions are not sufficient to characterize a valid 'oblique' syllogism in general. The following inference, for instance, even though satisfying all of the above conditions, is clearly invalid:

(1) John is the father of James;
(2) James is the father of Solomon;
(3) Therefore, John is the father of Solomon.

[75] Caramuel (1654: 406).   [76] Caramuel (1654: 406–7).

the premises and maintaining the same conclusion.[77] Consider, for example, the following 'Platonic syllogism':

(1) Every man is nobler than every horse;
(2) Every horse is nobler than some substance;
(3) Therefore, every man is nobler than some substance.

Inverting the premises, we obtain a new syllogism in which the expression 'every horse' becomes subject in the first premise and predicate in the second. Because in the 'Platonic syllogism' with inverted premises, 'every horse' plays a role analogous to that of the middle term in a first figure Aristotelian syllogism, Caramuel does not hesitate to recognize the new syllogism as belonging to the first figure.

Of the second figure, Caramuel gives, among others, the following instance:

(1) Every man is nobler than every lion;
(2) Every plane tree is not nobler than every lion;
(3) Therefore, every plane tree is not nobler than every man.[78]

And this can be reduced to the general, schematic form:

(1) Every X is nobler than Y;
(2) Every Z is not nobler than Y;
(3) Therefore, every Z is not nobler than X.

A third figure is derived from the second and in this case too Caramuel, as with the previous two figures, indulges in proposing a plethora of concrete instances. The inferences that Caramuel proposes are based, for the most part, as we have seen, on transitivity or on the composition of relations, but he seems not to be aware of this fact and he is more interested in displaying a list of concrete examples rather than in discovering the abstract form (or forms) of the displayed inferences.

Clearly, Caramuel's *oblique logic* was a step toward a logic of relations. Moreover, Caramuel's acknowledgment of the pervasive and preeminent role played by relations in the development of his ontological views was also an evident advance in this same direction. It is the comparison with De Morgan's essays, however, that puts back into perspective Caramuel's contribution to

---

[77] Caramuel (1654: 433).   [78] Caramuel (1654: 434).

the development of a logic of relations and clearly shows that Caramuel was still depending heavily on the scholastic tradition.[79]

4.4 Gerolamo Saccheri (1667–1733)

At this point, before concluding this chapter, let us discuss a passage from Saccheri's logic textbook, which is quite interesting for the present investigation. Gerolamo Saccheri, who is well known for his attempt to demonstrate Euclid's parallel postulate, wrote a text entitled *Logica demonstrativa* (*Demonstrative logic*), which was first published in 1697 (in Turin) under the name of Giovanni Francesco Caselette. After (at least) two new editions, each with the name of a different author, the *Logica* was finally published in 1701 under Saccheri's name. Unknown are the reasons (or reason) for this behavior on Saccheri's side.[80] Soon after its publication, Saccheri's *Logic* seems not to have had much fortune and it was only at the beginning of the twentieth century that it was rediscovered, thanks to a paper by the Italian philosopher Giovanni Vailati, who praised it as a very interesting and original piece of work.[81] Saccheri's *Logic*, indeed, is full of stimulating ideas, and it is clearly an outstanding text among the several logic textbooks of the eighteenth century.

In the *Logica demonstrativa*, Saccheri considers *oblique terms* according to the scholastic tradition. An oblique term is an expression that modifies the term to which it is attached, narrowing the range of its *supposition*. Thus, as we have seen, given a sentence like *Every horse of Socrates is running*, the scholastic logicians conceive the oblique 'of Socrates' as an expression that restricts the scope of *Every horse*. Saccheri classifies the oblique terms as *syncategorematic expressions*:

> Now, a term either signifies exactly by itself, so that by itself it can be a complete term of some proposition, as *Peter*, *man*, *white*, and thus it is called 'categorematic', or, on the contrary, because it does not signify in a sufficiently determinate way, it cannot play by itself the role of subject or predicate, but it can carry out only some tasks related to the subject and the predicate, specifying and modifying them. In this case, the term is called

[79] According to Dvorák, we may attribute to Caramuel the suggestion of employing as a rule for performing oblique inferences, the following 'general principle': "Things which are related to some third, are related themselves" (Dvorák (2008: 664). This principle, however, which corresponds to the algebraic composition of relations, is not explicitly stated by Caramuel.

[80] Cf., Pagli (2009).   [81] Vailati (1903).

'syncategorematic' and of this kind are prepositions, adverbs, conjunctions, oblique cases and partitive names [...][82]

As we have remarked above, and as Leibniz and many scholastic logicians had recognized before him, oblique terms imply the presence of relations; and we know that a typical difficulty of traditional Aristotelian logic was how to handle propositions containing relations and quantifiers. From this point of view, Saccheri is very interesting because in a 'Remark [*Adnotatio*]' to the 6th 'Proposition' of the fourth chapter on the *opposition of propositions*, he employs as an example a sentence containing an oblique term connected with two quantifiers. The passage is this:

> In the third place, consider that the propositions that above I stated to be contradictories must be different according to quantity not only for what concerns the right case of the subject, but even regarding its complements in the oblique case, if there are any. Therefore, the contradictory of the proposition *every horse of some man runs* should be *some horse of every man does not run*, which differs from the first not only for the quantity of *every horse*, which is the right case of the subject, but even for that of *some man*, which is its indirect complement in an oblique case.[83]

Here Saccheri considers a sentence in which a term ('horse'), which is within the scope of a *universal* quantifier is connected by means of the particle 'of' with another term ('man') that, in its turn, is within the scope of a *particular* quantifier:

(A)   Every horse of some man runs.

The particle 'of' in this proposition, corresponds to the genitive case in Latin and expresses a relation of possession, implying an *anaphoric* reference: there is a man *who* possesses horses and every horse *of this man* runs. Saccheri's claim concerning this proposition is that its contradictory is

(B)   Some horse of every man does not run.

Given a proposition p, its opposite contradictory is another proposition p^* that has a truth value opposite to that of p, so that if p is true, p^* is false and vice

[82] Saccheri (1701: 1).   [83] Saccheri (1697: 38).

versa: in other words, p and p^* cannot have the same truth-value. Now, this is what happens with A and B above. If B is true—if it is true that every man possesses some horse that does not run—then it is false that every horse possessed by some man is running (and vice versa). Clearly, Saccheri thinks of the contradictory of A as the *negation* of A:

(C)   Non: Every horse of some man runs;

and then he operates the switch of the quantifiers *every* and *some* that are under the effect of the negation, presupposing the 'law of double nega-tion' and that *every* is logically equivalent to *not some not* and *some* to *not every not*.[84]

This is a very promising result in view of a construction of a logic of relations analogous to that worked out by De Morgan and Peirce in the nineteenth century, but Saccheri did not show much interest in propositions in which relations and nested quantifications (and negation) are involved. Thus, in the entire *Logica demonstrativa* he leaves little room to the discus-sion of relational sentences and he does not consider any relational sentence, or any sentence with oblique terms more complex than that just examined.

Again, as in the case of Caramuel, Saccheri seems to be on the right path for developing a logic of relations but suddenly stops and does not cross the boundaries of traditional scholastic logic.

4.5   A First Conclusion

At this point, let us take stock. From the ontological point of view, most thinkers belonging to the scholastic and late scholastic tradition denied that polyadic expressions of the (mental or spoken) language denote any kind of

---

[84] Things, however, with sentence (A) are not so simple. (A), as suggested by an anonymous referee, is ambiguous and can be interpreted in two different ways:

(A1)   There is a man whose every horse runs.
(A2)   Every horse having a man as an owner runs.

In his Introduction to the Italian edition of the *Logica demonstrativa* (Saccheri 2012) Mugnai wrote that Saccheri should have chosen as the contradictory of (A) the sentence: (D) 'some horse of some man does not run.' Silvio Bozzi (Bozzi 2013), reviewing Saccheri (2012) remarked that (B) above is the proper contradictory of (A) and that Mugnai was wrong. Because Saccheri himself indicates (B) as the contradictory of (A), clearly, he interprets (A) according to the (A1)-reading and Mugnai is wrong (and Bozzi right).

reality 'in the external world.' Thus, the question naturally arises as to whether this *ontological* claim impeded the birth of a logic of relations analogous to that developed during the second half of the nineteenth century. It seems quite reasonable to think that an ontology attributing full-fledged reality 'in the outside world' to relations, understood as accidents with their legs in several subjects (to employ Leibniz's picturesque words), would have made easier the development of such a logic. Yet, a realistic stance of this kind was strongly in opposition to the dominant ontological framework of Aristotelian origin, according to which reality is composed only of individual substances and their accidents, which 'do not pass from subject to subject.'[85]

Paradoxical as it may appear, the possibilities for developing a logic of relations were more propitious, at least in principle, in the opposite ontological tradition represented by nominalism (or even 'conceptualism'). As we have seen, indeed, Peter Auriol and Leibniz, two authors quite sympathetic to nominalism, accepted without problem the existence of polyadic properties, even though confined to the 'realm' of mental objects. Once included in a domain of mental entities, relations, like numbers, could then be investigated according to their nature of pure 'logical things,' without any further ontological constraints. In this case too, however, we don't have the expected result. Thus, for instance, in the essays on rational grammar and in his logical writings, Leibniz is clearly puzzled by the 'new' relational inferences of Jungius (and Vagetius) and attempts to analyze them on the basis of 'more fundamental' logical principles.

As we have seen, Caramuel is the logician who, more than any other in the scholastic and late scholastic tradition, attempted to construct something analogous to a logic of relations. He is fully aware, for example, of the fact that the so-called *oblique terms* express relations and that among relations there are some that are *symmetrical*, whereas others are not. Moreover, he adopts a quite peculiar interpretation of the copula that reminds one of that which De Morgan was to propose in the second half of the nineteenth century. But Caramuel does not seem to be really concerned with mathematical or geometrical demonstrations, even though he explicitly says, presenting his *Logica obliqua*, that oblique inferences are of fundamental importance for doing mathematics.

As far as we know, it was mainly in Germany that, during the eighteenth century, after Jungius and Vagetius, the topic of oblique inferences was discussed in relation to the possibility of systematically using the traditional

---

[85] Cf., Thomas (1980: 206b, 246b) and Mugnai (1992: 35).

syllogism to carry out mathematical demonstrations. In the next chapter (Section 5.1) we will take up again the topic of oblique inferences, dealing with Andreas Rüdiger, an eighteenth-century German philosopher. We will contrast Rüdiger's positions with those of Christian Wolff, who instead seeks to bring obliquities back within the traditional syllogism. Then, we will broaden our gaze to consider the theses of Wolff himself, Rüdiger and other authors in the German milieu (Müller, Hoffmann, Crusius) regarding the relations between mathematics, philosophy, and syllogistics (Section 5.3). The reason why we dwell on Rüdiger is that, to a certain extent, his theses have interesting affinities with those to be argued by Kant.

5

The Extent of Syllogistic Reasoning

From Rüdiger to Wolff

5.1 Andreas Rüdiger (1673–1731) and His School on Oblique Inferences

As Jungius before him, Andreas Rüdiger, too, recognizes the existence of "inferences from the right to the oblique," even though he does not employ this expression to designate them. He claims, for instance, that all inferences analogous to the following one "cannot be reduced to syllogistic form":

> The right reason is from God; Therefore, he who despises the right reason despises God.[1]

Further on, in the same work, he proposes another example:

360 degrees make the entire circle,
180 degrees are the half of 360 degrees;
Therefore, 180 degrees are [equal to] the half of an entire circle.

This kind of syllogism, Rüdiger remarks, "has five terms and so it is not a syllogism at all, but a monster of syllogism. Yet, it is a true reasoning, even though not a syllogism."[2] As Rüdiger points out, mathematicians usually state *millions* of such inferences, which, being perfectly valid, "firmly reject all syllogistic rules."[3] Quite sharply, Rüdiger concludes that there is no way of accommodating mathematical reasoning to syllogistic laws.[4] We will discuss at length Rüdiger's position on the non-syllogistic nature of mathematics in Section 5.3.1 but we will first say something more about the discussion of oblique inferences in Rüdiger and his disciples.

---

[1] Rüdiger (1709 [1722]: 261). [2] Rüdiger (1709 [1722]: 290–1).
[3] Rüdiger (1709 [1722]: 291). [4] Rüdiger (1709 [1722]: 293).

Syllogistic Logic and Mathematical Proof. Paolo Mancosu and Massimo Mugnai, Oxford University Press.
© Paolo Mancosu and Massimo Mugnai 2023. DOI: 10.1093/oso/9780198876922.003.0006

Schepers had long ago emphasized the importance of non-syllogistic reasonings in Rüdiger and his school.[5] Rüdiger seems to have had no knowledge
of Jungius and to have arrived at his insights independently. Starting in 1709,
Rüdiger speaks of "ratiocinatio obiectiva" and treats it in most of his works.
One of the examples, already used in *Philosophia Synthetica* (1707), is the
following inference:[6]

No sin pleases God.
Thus: whoever justifies a sin, justifies something that does not please God.[7]

As Schepers observed, in Rüdiger's remaining works, the oblique inferences
are restricted to those in which the subject, expressed in the nominative case
becomes an object, thus expressed in the accusative case. For instance, in the
previous inference the Latin word *peccatum* (corresponding to 'sin' in the
premise) is in the nominative case but in the conclusion it is in the accusative
case. As pointed out by Schepers, this is not as general as Jungius' treatment in
which the oblique cases can be, in addition to the accusative case, in the
genitive, dative, and other cases.[8] On the other hand, Rüdiger also classifies
as non-syllogistic inferences those usually known as disjunctive syllogism,
modus ponens, and others.

A more extensive treatment of non-syllogistic inferences, and in particular
of 'ratiocinatio obiectiva,' is found in the second edition of *De Sensu Veri et
Falsi* (1722). The header of the chapters that we will list momentarily specifies
in Greek "asillogistos": "On sensual, or mathematical, ratiocination, non
syllogistic" (Book II, cap. IV); "On ideal ratiocination, by means of two
terms, non syllogistic" (Book II, chapter V). Chapters VI and VII are on
'syllogistic' reasoning by means of three and four terms, respectively. The
theory of syllogisms with four terms is original with Rüdiger.[9] Chapter VIII is
on "Ideal objective reasonings" and is again accompanied by the specification
"asillogistos." The same holds for chapter IX on "ratiocinatio transsumptiva"

---

[5] Schepers (1959).   [6] Rüdiger (1707: 58), Pars I, section III, Cap. IX.
[7] In the original Latin: "*Nullum peccatum placet Deo; Ergo: Quisquis peccatum excusat, aliquid
excusat quod Deo non placet.*"
[8] As pointed out in footnote 3 of Schepers (1959: 112), this contrasts with Wolff's treatment of
oblique inferences. Wolff is aware of the fuller set of oblique inferences treated by Jungius but thinks
that they can all be reduced to first figure syllogisms. See: *German Logic* (Wolff 1713), cap. 4, §29; for
the treatment of oblique inferences in Wolff's *Latin Logic* (Wolff 1735), see §§246, 247, 345, 365, 438.
We discuss Wolff on oblique inferences in Section 5.2.
[9] See Schepers (1959).

and chapter XI on "verbal or grammatical reasoning." Before providing more details let us also mention that even more chapters of *De Sensu Veri et Falsi* should be classified as dealing with non-syllogistic reasonings, as becomes evident from the later *Philosophia Pragmatica* (1723).[10] In it, four key chapters (Section I, Part I, Art. III, chapters 6, 7, 8, and 9) are devoted respectively to disjunctive, comparative, objective ratiocination, and causal and practical ratiocinations (these latter two were treated in *De Sensu Veri et Falsi* in Book II, chapters IX(B) and IX(C)). Once again, the header of each chapter specifies in Greek "asillogistos."

We will not analyze here in which sense Rüdiger thinks that all these forms of reasoning are non-syllogistic. The extent of his discussion would require too much space to be dealt with properly. In addition, since Rüdiger treats many of these non-syllogistic forms of reasoning independently of mathematics, a detailed treatment would also be superfluous for our immediate goals (this was not so with the medieval tradition we analyzed since it eventually led to Jungius and Vagetius who tied the issue of mathematical reasoning to oblique inferences). However, one should not get the impression that Rüdiger and his school belong to that trend of thinkers, represented by, among others, Bacon, Descartes, Tschirnhaus, Locke, and others, who denigrate syllogisms.[11] On the contrary, Rüdiger develops syllogistic theory in innovative ways and by means of his distinction between synthetic versus analytic uses of syllogistic inference he tried to account for the importance of syllogistic reasoning.[12] His main claim, and that of the thinkers he influenced, is simply that there are more valid inferences than just syllogisms. We will postpone to Chapter 6 a discussion of mathematical reasoning in Rüdiger and his claim that mathematical reasoning is non-syllogistic.

Rüdiger's student August Friedrich Müller treats non-syllogistic reasoning in his *Einleitung in die philosophischen Wissenschaften* (1733). Müller discusses inferences in chapters 14 to chapter 18 of the first volume of his *Einleitung*.[13] His treatment is derivative on Rüdiger and he, too, classifies certain sorts of inference, for instance those with one premise such as conversion and opposition, as non-syllogistic.[14] Confusingly, he does so in a chapter titled "Of syllogistic reasoning." A case of oblique reasoning, corresponding to

---

[10] In reality, *Philosophia Pragmatica* is only a later edition, with revised title, of the *Philosophia Synthetica*: see References.

[11] See, for instance, Boswell (1991: 126–30) and Petrus (1997: 15–22).

[12] See Schepers (1959) and Risse (1970).   [13] Müller (1733: 408–551).

[14] Müller (1733: 431).

what Rüdiger calls "objective ratiocination." is given under the labeling of "assumptive ratiocination in species." discussed in the same chapter, which he considers non-syllogistic:[15]

> The freedom of will is the basis of the entire theory of morals; thus, he who denies the freedom of will denies the basis of the entire theory of morals.[16]

An extensive discussion of this form of inference, this time with explicit reference to Rüdiger's "objective ratiocination," is carried out in section 25 of chapter 15. Among the examples, we find:

> The Apostles are God's messengers; he who listens to the Apostles listens to God's messengers.[17]

In chapter 15 (§26, §28, and §31), he adds disjunctive, comparative, and hypothetical reasoning to the class of non-syllogistic reasoning.[18] And just like Rüdiger, Müller considers mathematical, causal, practical, and verbal ratiocinations as non-syllogistic.

The emphasis on non-syllogistic reasoning is also very marked in Adolf Friedrich Hoffmann. Hoffmann's 1729 work against Wolff, *Gedancken über Wolffens Logik*, contains several sections on syllogisms aimed at debunking Wolff's claims that all inferential forms can be reduced to syllogisms of the first figure. Just like Rüdiger and Müller, Hoffmann excludes reasoning by conversion and opposition, causal reasoning, and practical reasoning from the realm of syllogistic reasoning. Wolff's mistake, he argues, rests on the prejudice that all inferences are syllogisms and this led him to the wrong definition of "ratiocination" (see pp. 20–1). The point is repeated in section XXI[19] where Hoffmann refers to Rüdiger's *De Sensu Veri et Falsi* (L.II, C.III, Schol. a) for a full refutation. In section XXIII, Hoffmann uses an arithmetical inference ("when 29 is subtracted from 38 what is left is 9") as a challenge to Wolff's claim that every inference can be reduced to a first figure syllogism.[20] He also claims that turning the inference into:

---

[15] See also Müller (1733: 494) for an explicit statement to this effect.
[16] Müller (1733: 455). Incidentally, Müller does not capitalize nouns in German.
[17] Müller (1733: 492). See also Schepers (1959: 113), who, however, cites from the first edition of Müller's work (1728: 433).
[18] Müller (1733: 494–6, 497–9, 501–4).   [19] Hoffmann (1729: 30–1).
[20] Hoffmann (1729: 32–3).

If 29 is subtracted from 38 then what remains is 9
But here or there, 29 is subtracted from 38
Thus, what remains is 9

is a mistake because in the first premise 29, 38, and 9 are taken *in abstracto*, whereas in the second premise and the conclusion they are taken *in concreto*, as specific instances of 29, 38, and 9 objects. In section XLI he also excludes disjunctive reasoning from the realm of syllogisms.[21] In section XLVI,[22] commenting on §427 of Wolff's *Logica Latina* where Wolff claimed the indispensability of syllogisms in demonstration, Hoffmann objects:

> For, if we had to reduce all demonstrations to syllogisms, they would have to be syllogisms either in virtue of their essence or in virtue of their external form. It is not possible that they should be so in virtue of their essence. For, as we have shown above again and again, not all types of inferences are *syllogistic*. Should they be so in virtue of their external form then I do not see the absolute necessity praised by the author.[23]

It is in the *Vernunftlehre* (1737) that Hoffmann systematically develops his thought on non-syllogistic reasoning and oblique inferences. His treatment is very extensive and contains some elements of novelty. While Aristotle was guilty of having restricted himself to only some valid forms of inference (which, according to Hoffmann, would be equivalent to writing a history of literature classifying all books according to their format) and Rüdiger had been able to improve (the first to do so, according to Hoffmann) the classification of inferences according to a better criterion,[24] Hoffmann suggests some improvements to Rüdiger's treatment, which he deems inadequate. In particular, he claims that all logical inferences are subsumption inferences but that this in no way implies that they are all syllogisms.[25] Among the interesting inferences discussed by Hoffmann we find the following relational inference:

> §578 When one sets a relation of a thing to another then one can infer the contrary relation of the latter to the former.[26]

---

[21] Hoffmann (1729: 66–7).   [22] Hoffmann (1729: 79–80).   [23] Hoffmann (1729: 79–80).
[24] Exploiting, according to Hoffmann, the type of abstraction that connects the ideas in propositions related as premises and conclusions.
[25] Hoffmann (1737: 533–4).   [26] Hoffmann (1737: 493).

Another inference is from the whole to the parts:

> §580 When one posits the whole, one posits at the same time all its parts. That is, from the assertion of the existence of a whole in a subject or in a space one must conclude to the assertion of the existence of all the parts in the same subject or space.[27]

The most interesting discussion in part I of his book consists in the extended analysis of the claim that every non-syllogistic inference can be turned into a syllogistic one.[28] He discusses at length the strategy consisting of turning the inferential rule into a premise and then proceeding syllogistically. As an example, he gives the following non-syllogistic inference:

The goal of war is peace [der Zweck des Krieges ist der Friede]

Peace is a good [der Friede ist ein gutes]

Thus, the goal of war is something good [Ergo ist der Zweck des Krieges etwas gutes].[29]

The syllogistic transformation is:

What has peace as its goal, its goal is something good [Was den Frieden zu seinem Zweck hat, dessen Zweck is etwas gutes]

War is something that has peace as its goal [der Krieg ist etwas, das den Frieden zu seinem Zweck hat]

Thus, the goal of war is something good [Ergo ist der Zweck des Krieges etwas gutes].

He grants that this can be done but that such a maneuver would, among other things, block the ability to find new truths.[30] Moreover, he objects that this is only achieved by turning an inference into a complex premise of the new argument. The distinction between rules of inference versus propositions capturing them is tantalizing but the discussion remains inconclusive.

---

[27] Hoffmann (1737: 494).   [28] Hoffmann (1737: 532–45).   [29] Hoffmann (1737: 537–8).
[30] Hoffmann (1737: 535).

5.2 Christian Wolff on Oblique Inferences

Christian Wolff (1679–1754) does not show very much interest in oblique inferences. At §246 of his *Latin Logic* (1735), for instance, he considers the case of a sentence with a universal quantifier ('a sign of universality [*signum universalitatis*]') in oblique form and suggests how to free the quantifier from obliquity. Given the sentence 'In every triangle the longest side subtends the greatest angle,' with the particle 'in' modifying 'every' [*in omni*], Wolff proposes to transform it into 'Every triangle has the longest side opposite to the greatest angle'; and the same holds for the sentence 'In some triangles, two angles taken together are greater than the third,' which becomes 'Some triangles have two angles, which taken together are greater than the third.'[31] Clearly, Wolff doesn't care about the use of the relative, which causes a substantial change of the form of the categorical sentences composing a traditional syllogism.

Wolff calls 'cryptic' the syllogisms with oblique expressions that conceal the role played by the quantifiers 'every' and 'some' in one (or both) of the premises. What Wolff seems to have in mind is a kind of syntactic rule, according to which the 'signs of quantity' must precede the other component parts of any sentence in which they occur. Once this requisite is satisfied, Wolff assumes a quite liberal attitude concerning the internal structure of a sentence, which, besides obliquities may contain anaphoric terms and other kinds of expressions traditionally not included in the categorical sentences composing a syllogism. This is in perfect agreement with his quite generic definition of a *syllogism* as "an utterance [*oratio*], by means of which a thought [*ratiocinium*] or a discourse is proposed in a distinct way."[32]

In §438 of the *Latin Logic*, Wolff proposes the following example of 'cryptic syllogism':

> That whole, whose part is equal to another whole, is greater than this latter.
> But *any part of a whole is equal to a part of this whole, i.e. it is equal to itself.*
> Therefore, *any whole is greater than its part.*[33]

He then makes manifest the inference hidden behind the 'cryptic syllogism' by means of a rephrasing which, in his eyes, should be an explicit, plain syllogism:

---

[31] Wolff (1735: 164, §246).   [32] Wolff (1735: 193, §332).

[33] Wolff (1735: 232, §438): "*Cuius pars alteri toti aequalis est, idipsum altero maius est. Sed quaelibet pars totius parti totius, hoc est, sibi ipsi aequalis est. Ergo totum qualibet sua parte maius est.*"

Everything whose part is equal to a whole is greater than this latter. But any
whole is a thing such that any part of it is equal to a part of the whole, i.e. to the
whole constituted by itself. Therefore, the whole is greater than any of its parts.[34]

This, however, is a complex inference involving anaphoric expressions and obliquities that can hardly be reduced to a 'classical' Aristotelian syllogism.

The notion of *equipollence* and the *principle of substitutivity* are the two basic tools that Wolff employs to show the intrinsic syllogistic nature of non-syllogistic inferences as, for example, the *inversion of relation*. To see in some detail how Wolff proceeds, consider the following sentences:

(1) 'Titius is Caius' father'; 'Caius is Titius' son.'
(2) 'The Sun lights up the building'; 'The building is lit up by the Sun.'

According to Wolff, the first two sentences are 'equipollent,' because they express 'the same complex notion,' that is, the same conceptual content.[35] Being equipollent, one can be substituted for the other, without changing the 'complex notion' corresponding to them. The same holds for the second pair of sentences above. As Wolff argues, because two equipollent sentences imply each other, we may assume one of them as *antecedent* and the other as *consequent* of a conditional, giving rise to an instance of what Wolff considers a hypothetical syllogism (an instance of *modus ponens*):

Of two equipollent propositions, if one is assumed as antecedent and the other
as consequent, you will get a hypothetical syllogism. The modus ponens of a
hypothetical syllogism [. . .] is: If the antecedent is [the case], the consequent
too is [the case]. But the antecedent is [the case]. Therefore, the consequent too
[is the case]. Indeed, if one of two equipollent propositions is granted, the
other must be granted as well. Because, if one of them is considered as the
antecedent and the other as the consequent, you will necessarily have a
hypothetical syllogism.[36]

Among the examples of *hypothetical syllogisms*, Wolff mentions:

If Titius is Caius' father, then Caius is Titius' son. But Titius is Caius' father.
Therefore, Caius is Titius' son.

[34] Wolff (1735: 232, §438).   [35] For the notion of *equipollentia*, cf., Wolff (1735: 175, §278).
[36] Wolff (1735: 234–5, §446).

Wolff considers the first premise of this instance of *modus ponens* ('If Titius is Caius' father, then Caius is Titius' son') as an *immediate consequence*; and he defines an *immediate consequence* as "that way of arguing, according to which, on the basis of logical reasons it is evident that, once a proposition has been granted, another proposition simultaneously follows."[37] Thus, Jungius' inversion of relation is for Wolff an *immediate consequence* but, unlike Jungius, Wolff doesn't consider it *non-syllogistic*. On the contrary, all *immediate consequences* are, according to him, *enthymematic inferences*, that is, syllogistic inferences in disguise: "it is evident that *all immediate consequences are enthymemes of hypothetical syllogisms, in all those cases, however, where the consequence is evident on the ground of logical reasons.*"[38]

To sum up: "the conclusion reached by means of an immediate consequence is inferred thanks to a categorical syllogism."[39]

In the *Latin logic*, Wolff explains this point, when discussing the immediate consequence of *modus ponens*. Here, he distinguishes two cases: (1) antecedent and consequent of the *modus ponens* have the same subject; (2) antecedent and consequent have a different subject. Let us consider case (1): doing this we will now avail ourselves of two schematic instances of, respectively, a *modus ponens* and a first figure syllogism (to simplify things, quantifiers are omitted, and the only verb employed is the copula 'is'):

If A is B, then A is C	B is C
A is B	A is B
Therefore, A is C	Therefore, A is C.

Discussing an analogous example, Wolff remarks that the two inferences have same conclusion and same minor premise. Moreover, the predicate (B) of the antecedent of the first premise of the *modus ponens* is the subject of the major premise of the syllogism; *and* the predicate (C) of the consequent of the first premise of *modus ponens* is the predicate of the major premise of the syllogism. Thus, Wolff concludes:

> Because in the *modus ponens* the minor [premise] is the antecedent and the conclusion the consequent, consequent and antecedent have the same subject (*by assumption*) and major [premise] and conclusion have the same subject, so they fit with the form of the first figure.[40]

[37] Wolff (1735: 238, §459).   [38] Wolff (1735: 238, §460).   [39] Wolff (1735: 239, §463).
[40] Wolff (1735: 221, §412).

In other words, making the suitable changes, "the hypothetical syllogism becomes a categorical syllogism belonging to the first figure."[41]

If case (2) holds, however, Wolff acknowledges that it becomes more difficult reducing *modus ponens* to a first figure syllogism.[42] In this case, he observes that to transform *modus ponens* into a first figure syllogism, one has to transform the second premise of the *modus ponens* into the middle term of the syllogism. Thus, given the following, concrete instance of a *modus ponens*:

If men sin, this world is not the best;
But men sin;
Therefore, this world is not the best,

Wolff proposes to transform it into:

No world, in which men sin, is the best;
This world is a world, in which men sin;
Therefore, this world is not the best.

Clearly, all these examples are ad hoc and quite artificial. In the case of a *modus ponens* with antecedent and consequent having the same subject, it is not difficult to see how the first premise can give rise to a universal categorical proposition. The conditional 'If Socrates is a man, then he is an animal' presupposes that every man is an animal. When antecedent and consequent are different, however, things are not so easy, as Wolff himself recognizes.

5.3 Mathematics, Philosophy, and Syllogistic Inferences in Wolff, Rüdiger, Müller, Hoffmann, and Crusius

5.3.1 Wolff: Every Mathematical Demonstration Is a Chain of Syllogisms

When we look at Wolff on the issue of the power of syllogistic reasoning, we witness a remarkable development. At the beginning of his career, under the influence of his teacher Neumann, and of Tschirnhaus, he denigrated syllogistic reasoning. The young Wolff shared with Neumann and Tschirnhaus an admiration for mathematics, which he studied intensively but mainly for methodological reasons: his plan was to extend its method to theology. He also shared

---

[41] Wolff (1735: 221, §412).   [42] Wolff (1735: 223, §415).

with them a disdain for syllogisms as unfruitful.[43] In 1705 Wolff started corresponding with Leibniz who expressed a different view than Wolff on the value of the syllogism. In an early work he had sent Leibniz, Wolff had claimed "Syllogism is not an instrument for finding the truth."[44] Leibniz, commenting on the essay, replied: "I do not dare say without further qualification that the syllogism is not an instrument for finding the truth."[45] This led Wolff to rethink his attitude toward syllogisms. In the *Ratio Praelectionum* of 1718 he recalls his early criticisms of syllogistic theory:

> Having learned that syllogisms are composed of the conclusion and the already given middle term, and given that no example was known in which from previously unknown premises one reaches a conclusion up to that point unknown, syllogisms seemed to me only a way to judge an already discovered truth but by no means a way to discover a hidden one.[46]

Later in the *Ratio* he embraces the idea that any conclusive demonstration can be reconstructed as a syllogistic one:

> [. . .] syllogisms are erroneously rejected by some moderns; one seeks for a criterion of truth in vain, for the precepts of the usual logic are sufficient to allow the examination of demonstrations through syllogisms.[47]

The position was not new with the *Ratio*. The first reference to the centrality of syllogisms in demonstration in Wolff occurs in the *Kurtze Unterricht von der mathematischen Lehrart* [*Short Tutorial on the Teaching Method of Mathematics*] which prefaces the *Anfangsgründe aller mathematischen Wissenschaften* [*Fundamentals of all the Mathematical Sciences*]:

> [. . .] No compelling demonstration can be carried out other than by following the way in which our thoughts follow one another according to syllogistic rules.[48]

---

[43] As we have seen, this was a very common position in the seventeenth century, shared by philosophers of many different dispositions.

[44] Wolff (1704: 289).

[45] Wolff (1860 [1963]: 17; see also 18). This episode is well known to Wolff scholars. See, among others, De Vleeschauwer (1932), Campo (1939), Arndt (1965), and Paccioni (2006).

[46] Wolff (1718 [1735; 1972]: 120 (Ratio, Cap. II, sec. II, §6)); see also Wolff (1841: 134–6).

[47] Wolff (1718 [1735; 1972]: 129); Wolff (1841: 137); see also Wolff (1718 [1735; 1972]: 129): "I then paid more attention to geometrical demonstrations and as soon as I did so I understood that if they are brought back to utmost precision they are composed of syllogisms connected among themselves in the same way in which, when I was a boy, I used to connect the proofs of the theses."

[48] Wolff (1710: §46).

In the previous section (§45) he explicitly states that in the mathematical sciences "everything will be inferred through the so called syllogism although at times, or rather most of the time, one leaves out one of the premises."[49]

The idea, already defended by Barrow, is that in their mathematical practice mathematicians proceed enthymematically,[50] that is, without always making explicit certain evident assumptions which are required for the completeness of the syllogistic reasoning.

The view that syllogism is central in the presentation of already established truths as well as for finding unknown truths is defended in detail in the so-called *German Logic* of 1713.[51] The major claims defended in section 4, "Von den Schlüssen [On Syllogisms]," which are relevant to our topic are the following:

(a) The first figure is sufficient for all syllogisms (§14);
(b) Everything that is discovered by human intellect is discovered through syllogisms; everything that can be demonstrated by argument is demonstrable through syllogisms (§20);
(c) Mathematicians often proceed by enthymematic reasoning (§21);
(d) A demonstration is a chain of syllogisms the first of which rests on definitions, clear experiences, and identical propositions (§21).

§22 is essential for capturing Wolff's position on the importance of syllogisms and merits to be cited *in extenso*:

Perhaps some will marvel at why I attribute so much importance to common syllogisms while they have become nowadays an object of derision on everyone's part. Let it be said that I am no admirer of antiquity and also not uninformed about the new discoveries; that moreover my teachers have inculcated in me contempt for common syllogisms and that I, just like the others, have laughed in my state of incomprehension; that however, after mature consideration, I have found the matter to be completely different [than I had at first believed] and that now, for truth's sake I am not afraid of defending, along with some great men, something that is considered as naive by those who have not penetrated deep enough into this matter.[52]

---

[49] Wolff (1710: §46).

[50] In the *German Logic* (Wolff 1713 [1727; 1965]) this claim is defended in §21 of chapter 4, "Von den Schlüssen". On the idea of enthymematic proofs in geometry, see Lassalle-Casanave and Panza (2018).

[51] On syllogism and invention, see also *German Metaphysics* (Wolff 1720; 1751; 1983, §345, §363). For a later theory connecting syllogisms and invention in mathematics, see the discussion of Salomon Maimon in Chikurel (2020) (especially §4.1.7.2).

[52] Wolff (1713: §22). The reference to "some great men" is quite obviously a reference to Leibniz who, through his letter to Wolff dated February 21, 1705, had influenced Wolff's attitude regarding the value of syllogisms.

After having rejected the idea that strict adherence to syllogistic presentation is the only mark of a perfect science, Wolff in the same section proceeded to claim:

(a) That in geometrical demonstrations one effectively thinks in syllogisms carried out in correct form;

(b) That in mathematics nothing can be found if not by means of syllogisms;

(c) That if in other disciplines one wants to prove and explain something according to mathematical method, it is syllogisms made in correct form that will lead us to the goal;

(d) That with the help of these syllogisms one can withstand the force of the most hidden errors.

The word used by Wolff is "Schluss" but the *Latin Logic* (1720) confirms that this is only a translation of "Syllogismus." In order to prove his claims, Wolff offered, in §23, a detailed syllogistic reconstruction of Euclid's proposition I.32 to the effect that the sum of the internal angles of a triangle equals two right angles. The proof is reconstructed as being made of four syllogisms and Wolff's analysis emphasizes that some premises in the proof are not explicitly stated because they correspond to information taken from the diagram or premises that are previously proven or clear from the context (he affirms, however, that all such premises are at least confusedly thought by the subject). This attention to the use of the diagram is noteworthy but unfortunately Wolff is, like most of his predecessors, oblivious to the problem of how to analyze the propositions occurring in these syllogisms so as to show that they fit the proper structure for syllogistic reasoning. Given the importance of Wolff's logic as one of the most authoritative sources for the idea that mathematics can be developed in syllogistic form, we translate most of §23.[53]

A geometrical proof takes place by means of syllogisms in form.

§23. In my Elements of Geometry (§101) one can find the following theorem with its proof. In every triangle ABC the sum of all its angles is 180 degrees.
    Proof.
    From the vertex C of the triangle draw a parallel DE to the basis AB. Let $1 = I$ and $2 = II$ (§97). Now, $I + 3 + II = 180$ degrees (§59). Thus, $1 + 3 + 2 = 180$ degrees.

---

[53] Wolff's proof of I.32 is also discussed in Anderson (2005); see also Anderson (2015).

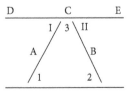

Whoever reflects on this proof, for instance, so that he is convinced by it, will think of nothing else but ordered syllogisms expressed according to their form. Indeed, from the cited result (§93) he assumes the following: all the alternate angles between parallel lines are equal to one another. The figure shows that I and 1 are alternate angles formed by the parallels AB and DE. Thus, he concludes that the angles I and 1 are equal to each other. One here already sees the first syllogism in form that must necessarily be thought if one wants to be convinced that angles I and 1 are equal to one another. In the same way he continues and assumes again from the cited result (§93) that all the alternate angles between parallel lines are equal to each other. Also in this case, the figure shows that II and 2 are alternate angles between the parallels AB and DE. Thus, he concludes again that the angles II and 2 are equal to each other. You see the second syllogism in form that anyone must think, if he wants to be convinced that the angles II and 2 are equal to one another. Moreover, from the cited result (§59) he assumes that all adjacent angles are equal to 180 degrees. The figure shows that the angles I, 3, and II are on a point C on the line DE. Thus, he concludes that the angles I, 3 and II taken together make 180 degrees. This is the third syllogism in form that one has to think to be convinced that the angles I, 3 and II taken together make 180 degrees. Finally, in order to be convinced that 1, 2, and 3 taken together make 180 degrees, the context draws to his mind that equal angles can be substituted for equal angles while preserving the magnitude of the latter. The proof shows that I and 1 and also 2 and II, are equal angles. Thus, he concludes that angle 1 can be substituted with I and 2 with II without altering the magnitude of angles I, 3 and II [taken together]. Then one also sees that 1, 2, and 3, make 180 degrees. This is the fourth syllogism in form that must be thought if one wants to be convinced that the angles 1, 2, and 3 make 180 degrees. In such a way the proof is made up of four syllogisms in form from which the premises are however omitted since they are recalled to the mind in part by the cited result, in part by the observation of the figure, and in part by the context.[54]

---

[54] Wolff (1713: 174–5, §23).

In §24, Wolff qualifies his claim that all mathematical truths can be found through syllogistic reasoning. He expresses doubt as to whether all recent developments in algebraic reasoning (probably thinking also of the differential calculus) can be so obtained but he certainly wants to claim that ordinary Euclidean geometry can all be discovered syllogistically as exemplified with the example of I.32. We will omit further details on this attempt because we are not so much concerned with syllogisms as tools for the finding of truths but mainly in whether syllogistic reasoning can recapture mathematical reasoning, independently of how the mathematical truths have first been found.

We conclude this section on Wolff with some passages from the *Elementa Matheseos Universae* (1713–15; itself a translation of the *Anfangsgründe* of 1710), for they reveal Wolff's knowledge of previous attempts at reducing mathematical reasoning to syllogistic reasoning. In §45, Wolff claims that the only justification in correct inference is given by the syllogistic forms studied in logic handbooks. In order to have a perfect demonstration all the premises of the syllogistic chain must be demonstrated by other syllogisms until one reaches premises which are either definitions or identical propositions. It is in §46 that he appeals to some of his predecessors as having established the possibility of syllogistic analysis of mathematical demonstrations:

§46. As a matter of fact, it would not be difficult to prove that a genuine demonstration capable of yielding full conviction can only occur if our thoughts are guided according to syllogistic rules. However, this is not the occasion to discuss this further. It will be sufficient to adduce some examples, as we are discussing a matter of fact. It is certainly not unknown that Clavius has resolved in syllogisms the demonstration of the first proposition of Euclid's *Elements*; and that, even more, Herlinus and Dasypodius have demonstrated, through syllogisms presented in a formal fashion, the first six books of Euclid's *Elements* while Henisch has done the same with the whole of arithmetic.[55]

While Herlinus and Dasypodius[56] are well known to us, the reference to Henisch is a novelty in the literature on this topic. Georg Henisch was the

---

[55] Wolff (1713–15: §45).

[56] In the *Kurtzer Unterricht von den vornehmster mathematischen Schriften*, Wolff writes that Herlinus and Dasypodius' book "die Beweise des *Euklidis* in förmliche Schlüsse zergliedert: welche Arbeit den Anfängern darzu dienen kan, daß sie begreifen lernen, wie aus Verbindung vieler Schlüsse ein gründlicher Beweis erwächset" ["it breaks up all demonstrations in Euclid into formal inferences; this work can be useful to the beginner so that he can learn to grasp how a grounded proof grows out of the connection of many inferences"] (Wolff 1731: 20, §38).

author of an *Arithmetica Perfecta et Demonstrata*.[57] However, a careful look at this treatise reveals that Wolff's attribution to Henisch of a presentation of arithmetical reasoning in syllogistic fashion is overly generous and ultimately unsubstantiated. Henisch does attempt at times to put arithmetical reasoning in the form of arguments with two premises and one conclusion but this only affects the presentation style and does not correspond to a serious engagement with the issue of syllogistic reduction. Thus, on this score, his treatise fares even worse than the one by Herlinus and Dasypodius. It is, however, worth remarking that the claims to reduce mathematical reasonings to syllogisms affected not only geometry but also arithmetic.

In §47, Wolff diagnosed why it has been so difficult for many people to recognize the syllogistic nature of mathematical reasoning. According to him, this is due to the use of figures and characters the observation of which, together with the use of other propositions, generally mislead us into not recognizing "how religiously the laws of syllogisms are preserved in mathematical demonstrations."[58]

Wolff is the great defender of the syllogistic nature of mathematical reasoning in the eighteenth century.[59] At the other side of the spectrum, though, we find Rüdiger and Reid. They will be discussed in Section 5.3.2 and Chapter 8, respectively.

5.3.2 Rüdiger and His School on the Non-Syllogistic Nature of Mathematics

Andreas Rüdiger is generally recognized as the first figure in the German Enlightenment to have carefully argued against the identification of philosophical and mathematical method. As such he counts as the originator of what might be considered an anti-Wolffian line.[60] This line, which includes August Friedrich Müller, Adolf Friedrich Hoffmann, and Christian August

---

[57] See Henisch (1609). In the *Kurtzer Unterricht von den vornehmster mathematischen Schriften*, Wolff says of Henisch's project that "er alle Beweise in förmliche Schlüsse zergliedert" ["he breaks up all demonstrations into formal inferences"] (Wolff 1731: 8, §12). Wolff gives Henisch's first name as "Gregor" instead of "Georg."

[58] See also German Metaphysics: Wolff (1720; 1751; 1983), §346, §349, §853, and Latin Logic: Wolff (1728), §545.

[59] Consider the following passage from Reyher (1693: cap. IV, §15) quoted in Basso (2004: 94, footnote 104), where we read: "Nobody should worry about Euclid's demonstrations on account of the fact that he does not prove his propositions by means of genuine and formal syllogisms, which is the true way of doing a demonstration. It can be enough here to employ sentences that virtually contain a syllogism, even though they do not express it explicitly."

[60] See Wundt (1945), Schepers (1959), Risse (1970), Ciafardone (1978), and De Felice (2008).

Crusius, is of great interest in connection to the development of Kant's early views. As is well known, the distinction between mathematical and philosophical method is found in Kant, starting at least from his 1764 work *Über die Deutlichkeit* (see Kant 1992b: 243–75). It has, however, been remarkably difficult to pinpoint with philological accuracy Kant's dependency on this tradition (with the exception of Crusius who is explicitly cited by Kant) and the interpreters who claim such dependency content themselves with postulating a dependency on account of similarity of doctrine.[61]

These thinkers are relevant to our topic on account of two major issues. The first, which we have already addressed in Section 5.1, is that they all argue for the existence of non-syllogistic forms of reasonings. The second issue concerns the distinction between philosophy and mathematics and the claim that mathematical inferences are not syllogistic in nature. While the two trends were connected in Vagetius, who gave an analysis of geometrical proofs that appealed to oblique inferences, this is not the case with the German thinkers we are considering who do not appeal to oblique inferences in clarifying the nature of mathematical reasoning. Rather, they argue for the non-syllogistic nature of mathematics from a variety of considerations. It is for this reason that we treated oblique inferences in Rüdiger, Müller, and Hoffmann in a previous section, independently of the topic of the non-syllogistic nature of mathematical proof. We now turn to the latter issue and discuss Rüdiger in Section 5.3.2.1 (and the Appendix 5.3.2.3) and his followers in Section 5.3.2.2.

5.3.2.1 Andreas Rüdiger on the Non-Syllogistic Nature of Mathematics

In giving a presentation of Rüdiger's position we also have to keep in mind the following. Rüdiger's revisions to his books indicate reactions to an evolving philosophical landscape, and this is particularly evident in the difference between the first and the second edition of *De Sensu Veri et Falsi* (1709, 1722). In 1713 Wolff published his *German Logic* that on many issues, and especially the one concerning philosophical and mathematical method, stands opposite to Rüdiger's viewpoint. Whether it is correct to see Rüdiger as the originator of an anti-Wolffian line is not easy to decide. It is possible that Rüdiger articulated the distinction between philosophical and mathematical method without knowledge of Wolff's early philosophy (Wolff's first book,

---

[61] For instance, in many of his books Ciafardone connects Kant to Rüdiger on mathematical definition (1978: 57; see also 76–7; see also De Felice 2016: 176) but nowhere do we find evidence that Kant had read Rüdiger or knew Rüdiger's theories. At the same time, it has hitherto escaped scholarly attention that Rüdiger was extensively discussed in books that Kant had read such as the logic treatises by Reusch, Corvin, and Knutzen.

Aerometria, which contains some reflections on mathematical method, came out in 1709). But it is also possible that Rüdiger's distinction between philosophical and mathematical method, which occurs for the first time in the 1709 book *De Sensu Veri et Falsi* and does not occur in the 1707 *Philosophia Synthetica*, might have been occasioned by Rüdiger's attendance of Wolff's lectures in Halle (where, according to Leichner 1727, Rüdiger was staying at the time).[62] The 1711 *Institutiones Eruditionis* (a second edition of *Philosophia Synthetica*) adds an additional chapter to the first edition of *Philosophia Synthetica* that reflects the change on Rüdiger's conception of mathematical method.[63]

Whether Wolff had read Rüdiger before publishing his *German Logic* is not known to us. Rüdiger is not cited in the *German Logic* (1713) or in the *Latin Logic* (1728). But Wolff surely must have known about him after Rüdiger's publication of *Christian Wolffens Meinung von dem Wesen der Seele und eines Geistes überhaupt; und D. Andreas Rüdigers Gegen-Meinung* [*Christian Wolff's Opinion on the Essence of the Soul and of a Mind in general; and Dr. Andreas Rüdiger's opposite Opinion*] (1727) and Hoffmann's publication of *Gedancken über Christian Wolffens Logic* [*Thoughts on Christian's Wolff's Logic*] (1729). There is also an attack on Rüdiger's views on mathematics by a follower of Wolff published in 1727 (Leichner 1727), whose second part attacks Rüdiger's position on philosophical and mathematical method).

For our goals, it will be sufficient to focus especially on the two editions of *De Sensu Veri et Falsi* (1709, 1722, Book II, chapters III and IV), and on

---

[62] Here is a quick summary of the loci in which Rüdiger treats the mathematical method:

1709, *De Sensu Veri et Falsi*, De Ratiocinatione sensuali, seu Mathematica, Asillogistos, Book II, chapter IV, §§1–11, pp. 204–10;

1711, *Institutiones Eruditionis*, Book I, Section II, Chapter VII [De veritate ratiocinativa mathematica sensuali seu mathematica], pp. 45–6;

1716, *Physica Divina*, Book I, Section I, §§40–59, pp. 13–21;

1717, *Institutiones Eruditionis*, De bonae definitionis requisite primario altero, distinctione idearum; Book. I; Treatise I, Part I, Section I, chapter VII, especially pp. 47–8; De Veritate Philosophica et Mathematica, Book I, Section II, Art. I, chapter IV, pp. 93–7; De veritate ratiocinativa mathematica sensuali, seu mathematica, Section II, Art. II, chapter II, pp. 103–7;

1722, *De Sensu Veri et Falsi*, De Ratiocinatione (a) sensuali, seu Mathematica, (b) Asillogistos, Book II, chapter IV, §§1–11, pp. 283–98;

1723, *Philosophia Pragmatica*, Section I, Part I, Art. III, chapter II, De Ratiocinatione Mathematica seu Sensuali, §§166–9, pp. 114–17;

1727, *Philosophia Pragmatica*, Section I, Part I, Art. III, chapter II, De Ratiocinatione Mathematica seu Sensuali, §§166–9, pp. 123–6.

[63] *Philosophia Synthetica* underwent several, heavily revised, editions but, confusingly, Rüdiger modified the title of his work: 1707, *Philosophia Synthetica*; 1711 and 1717: *Institutiones Eruditionis*; 1723 and 1727: *Philosophia Pragmatica*.

Physica Divina (1717, §40 et ff.), which are the most important for our topic. Since Rüdiger is a rather forgotten figure (he is known mainly to specialists of the German Enlightenment) and his Latin is challenging, we will devote to him more space than is warranted by his ultimate contribution to the debate. The reason for doing so is that this will provide our readers with a rather extensive selection of the texts where Rüdiger discusses mathematics and philosophy.

In his works, Rüdiger not only argues for the difference between mathematical and philosophical method but also provides several lines of argument in support of the claim that mathematical reasoning cannot be syllogistic. Let us focus on the relevant chapters of *De Sensu Veri et Falsi*, starting from the first edition (1709). We will be generous in quoting Rüdiger directly, for his tortured and idiosyncratic style defies easy summary.

In book II, section III (De veritate ratiocinativa in genere), Rüdiger distinguishes three modes of ratiocinative truth: sensual, ideal, and verbal.

> §1 What is properly said to be ratiocinative truth and how it differs from enunciative truth we have already shown above in chapter I, §1, 2, 3. Thus, we now only need to discuss the diversity of its modes. As in every reasoning three things are necessarily involved – *sensation, ideas* and *words*, – the intellect accordingly takes its start from these three principles to develop three different ways of reasoning. A difference which, although not at all discussed by other philosophers, it will be convenient to call *mathematical, ideal*, and *verbal*.[64]

Rüdiger restricts syllogistic reasoning to ideal ratiocination:

> §2 Since the doctrine hitherto accepted pretends to measure the vast field of reasoning with the narrow instrument of syllogism, it is easy to see how little it manages to encompass: not even a third of this field. Syllogism is in fact a type of ideal reasoning but two other kinds stand opposed to it, as it is evident from what we have argued so far.[65]

The thesis is innovative but strikingly paradoxical. In what sense are syllogisms not applicable to sensual reasoning? And why is mathematics restricted to sensual reasoning? Here is Rüdiger's explanation:

---

[64] Rüdiger (1709: 201, Book II, chapter III, §1).
[65] Rüdiger (1709: 202, Book II, chapter III, §2).

§3 Thus, for example, when mathematicians prove that *the three angles of a triangle are equal to two right ones,* they do not do so by means of ideas or through a syllogism but through *figures,* since the truth of the claim is grounded on certain *circumstances tied to sensibility,* which easily escape the untrained intellect. And while showing such figures, mathematicians claim they are *proving. Syllogism* is the best known example of ideal reasoning. Verbal reasoning occurs when the intellect infers, for instance, that *psychè* and *pneuma* are different from the fact that *psychekòs* differs from *pneumatikòs.*[66]

The contrast between ideal ratiocination and mathematical ratiocination is thus characterized along two different axes. First, mathematics deals with certain sensual circumstances while ideal reasoning does not. Moreover, syllogism is a form of ideal reasoning and, as we will see more clearly with the next citation, it has no role in mathematics:

§4 Ideal reasoning is either syllogistic or non-syllogistic (indeed, every non syllogistic reasoning is either tied to sensibility or it is verbal). If non syllogistic reasoning involves only two ideas, these ideas are either convertible or they are opposed to each other. Syllogistic reasoning, by contrast, *involves* either three ideas or *four,* but no more, as we will show at the appropriate junction. Thus, what philosophers have taught up to now with unanimous agreement, namely that *every syllogism consists of only and exactly three terms and three propositions,* is completely false.[67]

Rüdiger's thesis that there are syllogisms with four terms need not detain us here (see Schepers 1959 and Risse 1970).[68]

At this point, Rüdiger moves on to chapter IV devoted to an explication of "Sensual, or mathematical, ratiocination" which is characterized in the title itself as "asyllogistos."

---

[66] Rüdiger (1709: 202, Book II, chapter III, §3).

[67] Rüdiger (1709: 202, Book II, chapter III, §4).

[68] According to Rüdiger, the reason why such an important, but ultimately obvious, fact has been ignored is explained by the fact that syllogisms have only been considered analytically and not synthetically: "The reason why this has been hitherto ignored is that syllogisms have been considered by all *analytically* but not *synthetically.* Whoever has, however, considered syllogisms from a synthetic point of view and yet does not manage to appreciate this fact shows to be rather dense" Rüdiger (1709: 202, Book II, chapter III, §4). For reasons of space, we cannot enter into the details of this distinction.

Mathematicians and philosophers have failed to understand the difference between mathematics and philosophy. Philosophers have tried to foist syllogistic reasoning on mathematics while mathematicians have tried in vain to apply mathematical proof, which rests on the senses, even in ethics. But mathematical demonstrations and ideal demonstrations are of a different kind:

§1. This section can bring agreement between philosophers and mathematicians. Each group has hitherto pretended to impose stubbornly to the other group the form of reasoning that was familiar to it. Philosophers have tried to introduce syllogisms also in mathematics and mathematicians have in turn demanded proofs tied to sensibility even in moral theories. But neither group understood that mathematical proofs are diametrically opposed to the ideal ones and to the others.[69]

In what consists the sensual nature of mathematical demonstrations? In that the intellect grasps easily the sensual circumstances that constitute the situation described by the theorem. Rüdiger describes the process with Euclid I.47 (Pythagoras' theorem):

§2. Demonstration and mathematical reasoning are *tied to the senses* and consist in the following: *man understands and teaches certain circumstances tied to sensibility, which a less attentive intellect easily misses, and from whose gathering, however, emerges a certain truth.* Such circumstances tied to sensibility are found in Pythagoras' theorem, worthy of a sacrifice, that runs as follows. *In a right triangle the square on the greatest side, that opposed to the right angle, is equal to the sum of the squares of the remaining sides.* In such a case, many circumstances tied to sensibility offer themselves and this multifarious composition bothers the intellect to such a point that it cannot see the equality. By contrast, and this amounts to the same, if the medium square is described in such a way as to have a side in common with the smallest square, it will become clear that at this point there will remain only three circumstances, which, as soon as they become known, immediately lead to the conclusion, namely that in addition to the common parts what is left are two triangles of the greatest square. Thus, from the consideration of such circumstances tied to sensibility, the intellect persuades itself of the truth of Pythagoras' theorem.[70]

---

[69] Rüdiger (1709: 204, Book II, chapter IV, §1).
[70] Rüdiger (1709: 204–5, Book II, chapter IV, §2).

Even after rehearsing the proof of the theorem, it remains unclear why Rüdiger speaks of "sensual circumstances." The most plausible interpretation is that the proof is diagrammatic and it is the reliance on the diagram that constitutes the "sensual circumstances" which bring conviction to the intellect. Another important part of the example, the only theorem discussed in detail by Rüdiger in all his writing on mathematics, is the presence of part-whole reasoning. This will become a major criterion for distinguishing mathematical and philosophical reasoning.

From this Rüdiger draws a striking conclusion concerning the distinction between mathematical and ideal reasoning:

§3. From this example it becomes evident how distant *mathematical reasoning* is *from ideal reasoning*. Indeed, in the latter from one proposition one obtains another proposition but in the former many propositions yield a single proposition. The plurality of those individual circumstances tied to sensibility yield a single conclusion. In ideal reasoning, by contrast, we argue by *assumptio* or *transsumptio*: in this case no new idea is assumed while through the conjunction of the plurality of circumstances tied to sensibility emerges a single conclusion.[71]

Rüdiger's reasoning here seems unpersuasive. While it is possible to accept that in Pythagoras' theorem we use several previously established facts to reach the conclusion, this hardly shows that only a single conclusion can be drawn from those facts. Other conclusions could be drawn from the same facts. Moreover, one could combine into a single sentence all the "circumstances" leading to Pythagoras's theorem. Conversely, short of giving a very special definition of ideal reasoning, it is unclear why ideal reasoning might not in fact rely on several circumstances, expressed in various premises. Finally, it is totally unpersuasive that no new idea is assumed in the geometrical reasoning; and, in any case, no argument is given why this property would be generalizable even if it were true of the proof of Pythagoras' theorem. But, let's continue reading:

Ideal reasoning can be constructed as a syllogism if a new idea is assumed or one appeals to an already existing one. Mathematical reasoning, by contrast, does not admit syllogistic laws. As a rule, ideal reasoning does not allow

---

[71] Rüdiger (1709: 205–6, Book II, chapter IV, §3).

more than four terms while mathematical reasoning is not limited by any definite number of terms.[72]

Rüdiger contrasts ideal and mathematical reasonings by claiming that syllogistic reasoning is the proper logical scaffolding of ideal reasoning whereas mathematics does not admit such reasoning. Moreover, whereas ideal reasoning does not go beyond four terms (consistently with Rüdiger's claim that syllogisms can have four terms), there is no limit to the number of terms that can appear in mathematical reasoning.

In section 4, Rüdiger diagnoses what has gone wrong with the attempts of the "inept" Descartes and of the "insane" Spinoza to draw ideal conclusions from mathematical reasoning:

> §4. From this it becomes evident that mathematical reasoning *agrees* with ideal reasoning in that the former, just as the latter, proceeds through statements and obtains them through definition and division. This makes it even clearer that it is not possible to know mathematical facts with ideal reasoning but also that it is not possible to investigate moral subjects by means of mathematical reasoning. Thus, Descartes and his followers proceed in vain when they try to demonstrate things pertaining to moral subjects by mathematical means. And Spinoza is crazy when he attempts to demonstrate his atheism by means of the mathematical method. Everyone should remain within his own boundary.[73]

According to Rüdiger, the distinction between mathematical method and ideal (philosophical) method shows that it is vain to try to prove philosophical claims with mathematical reasonings and just as hopeless to try to prove mathematical results with philosophical arguments. The root of Spinoza's atheism rests on this methodological confusion.[74]

Rüdiger now describes the two main branches of mathematics in order to be more precise on their reliance on sensual reasoning:

> §5. Mathematical reasoning can be divided into *geometrical* and *arithmetical*. And this division concerns not the difference in reasoning itself but rather

---

[72] Rüdiger (1709: 205–6, Book II, chapter IV, §3).

[73] Rüdiger (1709: 206, Book II, chapter IV, §4).

[74] The reference to Descartes and Spinoza in the context of the debates concerning the application of the mathematical method in philosophy is a constant topic in eighteenth-century philosophy (see Basso 2004: chapter 2).

the object which is roughly twofold: *magnitude* and *number*. Indeed, in both cases the object concerns the circumstances related to sensibility and in both cases a manifold conjunction yields the conclusion. Every number, in fact, consists of units that if connected in various ways show the arithmetical truth from the point of view of the senses. If, for instance, three units generate nine units it is then certain that six units generate eighteen units.[75]

Here it would seem that the sensual circumstances which are at the basis of arithmetical reasoning consist in the fact that counting proceeds with units that, when variously connected, display the sensual nature of reasoning. As we will see when commenting on the second edition, Rüdiger later refines his thought by articulating a formalistic account of mathematical reasoning based on concrete symbols.

What we have said so far covers what Rüdiger says in the first edition of *De Sensu Veri et Falsi* concerning the non-syllogistic nature of mathematics. He has much more to say about the distinction between mathematics and philosophy, especially on the nature of definitions, but as this is independent of the issue of the use of syllogism in mathematics, we will not discuss this part of his thought any further.

The non-syllogistic nature of mathematics remains an invariant in Rüdiger's thought even though new elements appear in successive works. For instance, in *Physica Divina* (1716), which constitutes the transition point to the second edition of *De Sensu Veri et Falsi* (henceforth *DSVF*), we find an interesting contraposition between philosophy and mathematics based on the claim that philosophy proceeds using distinctions between genus and species (hence syllogistically) whereas mathematics proceeds according to part-whole analysis:

§45.III. The *divisions* of the philosophers are usually given in terms of *genus* and *species* while those of mathematicians in terms of *whole* and *parts*. These two types of division differ greatly. Indeed, the genus is wholly in any species, the whole by contrast is not at all in any part. Thus, the species can be subsumed by the genus but not the part by the whole and the latter two can be compared by means of the usual algebraic *plus* and *minus*. By contrast, philosophy, if it happens to be treating the whole, does it by means of predication not by means of computation. This difference can also be shown a priori as follows. A *possible* object *necessarily has parts but does*

---

[75] Rüdiger (1709: 206–7, Book II, chapter IV, §5).

not necessarily have a species because in the case in which the inferior things agree there is no diversity and where there is no diversity there is no species. Consequently, to any possible object one can always apply mathematical reasoning by means of parts but not philosophical reasoning by means of species. Finally, mathematical division (since mathematicians divide in order to be able to compute) proceeds correctly and rightly to *infinity* or, at least, to the indefinite. Philosophical reasoning, in truth, since it divides with the aim of understanding the nature of the divided thing never reaches *beyond the physical principles.*[76]

We now move to the second edition of *DSVF* (1722).[77] In Book II, chapter III, Rüdiger speaks of ratiocinative truth. A threefold distinction (ideal, sensual, verbal) is obtained by appeal to the observation that sensations, ideas, and words are involved in every reasoning. Ideal ratiocinative truth pertains to philosophy and is distinguished in metaphysical (existential) and disciplinary (essential). The latter in turn into causal (physics) and moral (practical). That gives five ways of reasoning with a further distinction introduced by bringing in matter and form:

Since Aristotle's time up to now every learned person has been convinced that reasoning is uniform in all branches of knowledge and that it has syllogistic form. I will by contrast easily prove the opposite to all intelligent men who are concerned with truth if I will manage to show that these five types of reasoning differ completely according to both *matter* and *form*, a difference greater than which, concerning reasoning, cannot be conceived. Indeed, as it will be shown more amply later, the *matter* of grammatical reasoning are the names and the verbs while the *form of the argumentation* is given by their inflection according to the different parts of discourse. The *matter* of *mathematical* reasoning are quantities and the theorems composed by these; the *form* is the enumeration (measurement too, if one wills) of such quantities. The *matter* of *metaphysical* reasoning are the categories and the opposites and the relevant theorems; its *form* consists in the subordination of predicables and the distinction of opposites. The matter of *causal* reasoning are the causes and causal theorems while its *form* is the *demonstration of the effects* from the power of causes. Finally, the matter of practical reasoning are *the ends* and the theorems whose predicates express a determinate end; the

---

[76] Rüdiger (1716: 15–16, Book I, chapter I, Sec. I, §45).

[77] We will try to be brief when the material is identical to that of the first edition.

form is the *demonstration of the means* to order to achieve the end. Since all these matters and forms differ *toto caelo*, as one says, one should be ashamed to keep humming the same old nursery rhymes, that every form of argument is syllogistic.[78]

And in §4:

As far as I am concerned I will not deny that syllogisms can be divided into these single modes despite the fact that these modes are completely different among themselves. But even he who will pay scant attention to the matter will see that they are strange and deceitful since they introduce by force the existence of the thing when, with respect to it, there subsists no doubt or question. By contrast, they do not convey anything at all with respect to the essence – be it physical or moral or mathematical, – which is never unveiled through syllogism and that no Daedalus of syllogism in any century has ever managed to unveil, as it will result clearer than the midday sun when considering mathematical reasoning as we will do in the next section when we will treat Pythagoras' theorem. In truth, syllogisms only accomplish the following: if a theorem is obtained, be it mathematical, physical, or moral, on the basis of pure reasoning and without use of syllogisms, syllogisms remind solemnly of what everyone who had a grasp of the meaning of the theorem knew from the outset, namely that in that argument there occurred a theorem. By contrast, the mechanism of reasoning consisted not in the application of the theorem, which is typical of ordinary knowing and that has no use for the needless tool of syllogism, but in the discovery of the theorem itself.[79]

In the following sentences, Rüdiger repeats *verbatim* some of the claims that had already been made in the first edition of *DSVF* concerning the fact that mathematicians do not proceed using ideas or syllogisms but rather through figures which ground the sensual circumstances yielding their theorems.[80]

Ideal ratiocination, that is, philosophical reasoning, can be either syllogistic or non-syllogistic. By contrast, verbal and mathematical reasoning is all non-syllogistic. Non-syllogistic forms of reasoning are given by premises with only two terms [ideas] (as it happens in conversion and opposition) or with more

---

[78] Rüdiger (1722: 279–80, Book II, chapter III, §1, note §3).
[79] Rüdiger (1722: 280, Book II, chapter III, §IV).
[80] In some of the following we need to go over some material that overlaps with that treated in the first edition.

than four terms [ideas]. Rüdiger claims that syllogisms can have either three or four terms and that, contrary to the common agreement of philosophers, it is absolutely false that syllogisms consist of exactly three terms and three propositions.[81]

We now move to discuss Chapter IV whose title is: *De Ratiocinatione (a) sensuali, seu mathematica, (b) asillogistos*. The (a) and (b) appearing in the title refer to two notes that we need to discuss first. We start with note (a).

According to Rüdiger's empiricism, all knowledge starts with sensory perception. Mathematics is no exception but it is not on this account that it is said to be sensual. Rather, it is the form of reasoning that is based on sensual experience.[82] Rüdiger argues as follows. Pure mathematics is the basis of impure mathematics (by which he means applied mathematics). Pure mathematics consists of arithmetic and geometry. All of mathematics proceeds from arithmetic (which Rüdiger identifies with algebra) and geometry. But arguments in geometry, as they measure quantities, are not different from arithmetical arguments and thus all of mathematics, when looked at from the point of view of reasoning, reduces to arithmetical reasoning, that is, the four basic operations (addition, subtraction, multiplication, division). But each such operation is an enumeration [*numeratio*] or computation of concrete items insofar as the terms of the computation are given by the senses. Thus, the nature of mathematical argumentation is sensual and since none of the other forms of argumentation shares the same form of argumentation with mathematics, mathematical argumentation is the only one of the forms listed by Rüdiger to have a sensual basis. Rüdiger grants that the object of mathematics can be either concrete [sensual] or abstract and that his claim is restricted to the form of reasoning:

All of pure mathematics is the basis of impure mathematics and thus all the parts of mathematics are based on arithmetic and geometry. But the way of arguing in geometry is nothing else than the *arithmetical* one since quantities are also measured by means of numbers. It follows that *all of mathematics*, when considered with respect to its arguments, is correctly reduced to arithmetic. Moreover, all of arithmetic, as well as *algebra,* is founded on *addition, subtraction, multiplication* and *division*; every addition,

---

[81] Rüdiger (1722: 280–1, Book II, chapter III, §IV).

[82] "Every reasoning is, with respect to its origin, no doubt based on the senses. Thus, it is not from this point of view that mathematics is said to be tied to sensibility but on account of its way of inferring, which only in this type of reasoning has a sensual origin" Rüdiger (1722: 283, Book II, chapter IV, note (a), §1).

multiplication, subtraction and division is founded on *enumeration* and every enumeration concerns individuals *since their boundaries are perceived with the senses*; nay, such boundaries are the principles of enumeration, that is they are the true and real units. Thus, every enumeration is tied to the senses. After all, the entire mode of reasoning in mathematics is enumeration and thus this mode of reasoning in its entirety is tied to the senses.[83]

Since this captures the essence of mathematical reasoning and no other form of reasoning rests on enumeration, Rüdiger concludes that only mathematics is sensual:

> The mode of reasoning, on the other hand, is the form, i.e. the nature and the essence of reasoning. And since no other type of reasoning is founded on enumeration (see previous chapter, observation (a) § 2) no other type of reasoning is tied to sensibility and thus only mathematics is tied to the senses; hence it differs maximally from the other species of reasoning.[84]

Faced with the objection that mathematics refers to universals, or *abstracta*, and not only to individuals, Rüdiger retorts:

> §2. It could happen that those who would prefer this claim to be false might object that they do not only deal with individuals but also, and above all, with universals. This, however, I do not deny. Indeed, I do not speak about the objects but only of the mode of reasoning and of drawing consequences, which mathematicians can apply only enumerating and which results sufficiently from plus, minus and equal in algebra.[85]

One must pause here and ask how innovative this position is. It strikingly reminds us of Hilbert's conception of metamathematics where one deals with the concrete signs, independently of what they might refer to. But even without going that far, the emphasis on mathematics being based on signs is also found in Kant's essay *Über die Deutlichkeit*. As far as we can tell, this position is original with Rüdiger and it might well be an indicator of Rüdiger's influence on Kant. But lacking any direct evidence we will leave it at that. There is of course much to object to the quick identification of arithmetical

---

[83] Rüdiger (1722: 283, Book II, chapter IV, note (a)).
[84] Rüdiger (1722: 283, Book II, chapter IV, note (a), §1).
[85] Rüdiger (1722: 283, Book II, chapter IV, note (a), §2).

and geometrical reasoning but here we are more concerned to describe Rüdiger's position, which strikes us as historically interesting, than to evaluate its argumentative strength, which we admittedly find very weak.

What we just discussed corresponds to note (a) as indicated in the title of Rüdiger's chapter. Note (b) in Rüdiger's chapter briefly claims that the non-syllogistic nature of mathematics results a priori from what has been said in the previous chapter and *a posteriori* from an example (the Pytagorean theorem) to be discussed below. We now go back to the main text. In §1 Rüdiger repeats *verbatim* what had been said in the first edition, namely that mathematicians and philosophers try in vain to foist on each other their method. §2 is also quite similar to the first edition except for the novel reference to "computation" when discussing Pythagoras' theorem. The latter is given in more technical terms than it was done in the first edition but there is no need to repeat it here. In §3, the main text between the first and the second edition is almost identical except that when contrasting mathematical and ideal reasoning, Rüdiger adds "syllogistic" to characterize ideal reasoning. And speaking of the conclusion of a mathematical argument he specifies that it emerges by enumerating (*numerando*) not by subsumption:

> From this example it will be evident how much (c) this *mathematical* (!) reasoning differs from the *ideal* syllogistic one. Indeed, in the latter from a single proposition one infers another proposition; in the former, however, from many propositions one infers a single proposition. In fact, a plurality of single circumstances tied to the senses yields a single conclusion. In the case of ideal reasoning we argue by means of *assumptio* or *transsumptio* of an idea; by contrast, in mathematical reasoning no new idea is assumed and the conclusion is obtained through the conjunction of a plurality of circum-stances tied to the senses by virtue of an enumeration and not by means of a subsumption (I am referring in this case to mathematics as such). And from this it is sufficiently clear that such mathematical arguments (d) do not admit at all the laws of syllogism.[86]

However, Rüdiger also adds the following in contrasting mathematics with ideal physical reasoning and moral reasoning:

> From this example it can become clear in what way mathematical reasoning differs

---

[86] Rüdiger (1722: 285–6, Book II, chapter IV, §III).

(1) From ideal syllogistic reasoning. In the latter, indeed, from a proposition one infers another one, while in the former from many propositions one infers a single one. In such a case, a plurality of sensual circumstances yields a single conclusion.

(2) From ideal physics, since it does not define either the species, or the type of action, or the cause or the physical reasoning, while instead it assumes many false causes or modes, as the illustrious mathematician Isaac Newton himself concedes (Phi. Nat. Math. Def. VIII, p. m. 5).

(3) From ideal moral theory, since it does not treat in depth the ends and means by genuses, as it is proper of moral reasons, and treats, by contrast, the ends from the quantitative point of view and of the means similar to these.[87]

The real novelty, with respect to the first edition of *DSVF*, consists of two long notes (c) and (d). Note (c) is divided into four main sections. The first refers to *Physica Divina* and brings out the importance of the notion of possibility. Since it is not central to the topic of the non-syllogistic nature of mathematical reasoning we will not discuss it here. By contrast, the long Note (d) expands and develops the issue of the non-syllogistic nature of mathematical reasoning. On account of its length and for those who find Rüdiger's position of interest, we treat Note (d) in an appendix to this chapter (Section 5.3.2.3). The reader who is not especially interested in Rüdiger can skip the appendix, for what we have said so far about Rüdiger is enough to convey the general gist of his position. A full analysis of Rüdiger's comparison between mathematics and philosophy would have to proceed to sections 4 to 11 of his text but what we have discussed (together with the discussion of Note (d) in 5.3.2.3) fully covers what he says on the non-syllogistic nature of mathematical reasoning.

There is no denying that Rüdiger's arguments are often unclear and leave much to be desired. We have decided not to press him on the cogency of his assumptions and argumentative line as we feel that the exercise would not be rewarding. Rather, we want to emphasize the historical importance of his position for the topic we are covering. Rüdiger argued for the non-syllogistic nature of mathematical reasoning not from a detailed analysis of mathematical proofs but rather from an ontological position (namely, the ontic structure of predication accounts for metaphysical truths whereas a different type of ontic relation, based for instance on part-whole, is a feature of many mathematical proofs).

---

[87] Rüdiger (1722: 286–7, Book II, chapter IV, §III).

With this we have fully covered Rüdiger's thought on the non-syllogistic nature of mathematical reasoning. In *Philosophia Pragmatica* (1723, 1727) he treated the matter in a more concise way without adding any new elements to his conception.

5.3.2.2 Syllogism and Mathematical Reasoning in Müller, Hoffmann, and Crusius

We have seen that Rüdiger's position on the difference between philosophical and mathematical method appeals to a variety of aspects such as the nature of philosophical and mathematical reasoning, definitions in mathematics and philosophy, and so on. Many of these features are also present in Müller's discussion. However, here we will focus on Müller's arguments aimed at showing that mathematical proofs are non-syllogistic. Müller is somewhat more systematic than Rüdiger in articulating the position and thus by summarizing his presentation we shed some further light on the position.

The two most substantial discussions of mathematics in Müller are in chapter 8, section 19, and chapter 14, section 9, of his *Einleitung in die philosophischen Wissenschaften* [*Introduction to Philosophical Sciences*] (1728, 1733).

In chapter 8, section 19, Müller states that mathematical *abstracta* are the grounds for numbering and measuring quantities. The art of finding mathematical propositions through abstract computations is algebra, which Müller characterizes as the special logic of mathematics. But through numbering and measuring only quantities can be found and thus according to Müller algebra is only applicable in the mathematical disciplines. Müller is here arguing against Wolff (*Anfangsgründe der Algebra* [*First Principles of Algebra*], §§11, 12) who had argued for the universal nature of algebra on account of the fact that there is nothing in the created nature that lacks quantity or measure. Müller observes that mathematicians must comply with general methodological rules (the starting points should be true, objects must be properly defined, the reasoning must be valid, and so on) but that these methodological requirements are not proper to mathematics. Against Wolff he claims that it is improper to call such methodological requirements "the mathematical method," for they were neither discovered by mathematicians nor are they exclusively used by mathematicians. After a long discussion, Müller concludes by reminding the reader that the method of thinking rigorously ("die methode gründlich nachzudenken") is not exclusive to the mathematical disciplines or to mathematicians.

The key issue for Müller is the type of connection [*Zusammenhang*] that one finds between mathematical truths, both between their subjects and predicates and between the premises and conclusion of a mathematical argument. This connection is altogether different from the connection of metaphysical, physical and moral truths. And this accounts for why the rules of discovery and resolution of mathematical truths are altogether different from the rules that are appropriate for the other realms.

Müller's theory, just like Rüdiger's, rests on the distinction between the types of *abstracta* that are connected through valid inferences. There are two major categories: existential inferences and causal inference. The former type of inference is so called from the fact that the consequence of the inference rests on the relation of "abstractorum existentialium," namely the relations of genus, species, difference, and accident:

> These inferences are those in which from the fact that the predicate of the basic proposition belongs or not to the subject, it is inferred that therefore also the abstract *existentialia* of the predicate must belong or not to the subject and the predicate must belong or not to those abstract *existentialia* of the subject. Therefore, through the existential inferences no other than existential truths will be found, that is such that can be predicated from the abstract *existentialia* of the subject that one contemplates according to their mere existence in the subject without for the moment seeing their original causes.[88]

While this also is the basis for distinguishing existential from causal and moral inferences, let us focus on mathematical and existential inferences. Let us use an example given by Müller to try to understand the nature of existential propositions and inferences. Müller offers the following example:

> All infractions are acts undertaken freewillingly against the law.[89]

He calls this an existential proposition "for it shows what must be present or should be present in an act if it is to count as an infraction." From that proposition one can infer, through existential inferences, other existential truths. That is, in as far as the predicate exists in the subject, it follows that other abstracts that exist in the predicate must exist in the subject; conversely, that which is opposed to the predicate is also opposed to the subject. It would follow in this way that a child who does not yet have the use of reason cannot commit an infraction (for his act could not be committed with free will). It is

---

[88] Müller (1733: 428; chapter 15, §1).   [89] Müller (1733: 429; chapter 15, §1).

not hard to see that these instructions correspond to the *dictum de omni et nullo* and indeed Müller claims that the syllogistic inferences are the most important existential inferences.

In contrast to the above, Müller claims that mathematical reasonings have nothing to do with existential truths.

> That is, mathematical inferences rest on the different relations of quantities and the latter on counting and measuring. And thus a mathematical infer-ence is, according to its form, nothing else than a computation [rechnung]. In the same way its correctness, connectedness or connection, rests merely on the rules of calculation. For instance, two and three together make five; five and four make nine; consequently, two and three and four of necessity make nine [...] The existential inferences, by contrast, rest on a completely different kind of relation, namely on the relation of the *abstractorum exis-tentialium* or *predicabilium* to one another, whose basic rules have been given in chapter 8, section 21, and the latter have nothing to do with the calculation of quantities and thus with mathematical inferences. For instance, all conditions that consist in duties of human beings towards human beings are worldly conditions: consequently, the condition of being married is a worldly condition (chapt. 9, §21 reg. 2). Who would want to claim that the form or consequence of this existential inference could be justified in the same way as the form of that mathematical inference from the rules of computation? or that the consequence of that mathematical infer-ence could be justified in the same way as the consequence of this existential inference on the basic rules of *praedicabilia*? Neither one works. For, where there are no numbers or quantities, there cannot be computations. Therefore, no inference that is not, unlike mathematical ones, a computa-tion, can rest on a computational rule nor can its consequence be considered as a mathematical consequence. In the same way, where there are no *genera, differentiae specificae* and *accidentia*, there cannot be the rules of these *pradicabilia* on which the consequences of all existential inferences rest.[90]

And now follows the argument that excludes that computations can be reduced to existential inferences:

> In all true mathematical inferences one calculates; nothing can be calculated but numbers: no single number relates to any other as a *genus*, a *differentia*

[90] Müller (1733: 423–4; chapter 14, §11).

specifica or an accidens of the same; thus, the relation of *genus*, of *differentia* or *accidens* never takes place in any calculation and thus in no mathematical inference; and therefore no mathematical conclusion, or computed sum, as it follows from its premises can be justified from the rules of *pradicabilia* [ch. 8, sect. 21].[91]

Müller concludes to the complete heterogeneity of existential and mathematical inferences. It is to be emphasized here that when Müller speaks of mathematical computations he does not have in mind something like an axiomatized arithmetical system but rather the addition rules that are used for adding, say, 36, 45, and 19 and that instruct us to add 6, 5, and 9, write 0 and carry over 2 in the next column, and so on. In addition to the obvious inadequacy of this process as an analysis of arithmetical inferences, how this is supposed to account for geometry remains a mystery.

Müller also claims that computing the sum of 36, 45, and 19 yields 100 and since this is the result of a computation the argument and the conclusion are mathematical and in no way syllogistic. However, when we apply this computation to specific objects, say talers, we infer that 36 talers plus 45 talers plus 19 talers yield 100 talers. According to Müller, this conclusion is obtained by subsumption from the first mathematical truth, and the argument is syllogistic.

Müller is then also forced to claim that the following inference, cited by others as an example of a syllogistic mathematical argument, is syllogistic but not mathematical:

The sum of the angles of every triangle is equal to two right angles

Every scalene [triangle] is a triangle

Thus, the sum of the angles of every scalene triangle is equal to two right angles.[92]

Contrary to the absolutism of those who think that every type of inference is syllogistic or that every kind of inference is algebraic, for example, Müller takes the position that there are a variety of heterogeneous inferential rules that depend on the connection between the *abstracta* related in the proposition and inferences. His diagnosis is that one often gets misled by the fact that these inferences often appear with an "Atqui" ['and'] connecting the premises and

---

[91] Müller (1733: 424; chapter 14, §11).   [92] Müller (1733: 426; chapter 14, §11).

an "Ergo" ['therefore'] leading to the conclusion so that this exterior form tempts people into seeing in them syllogisms (as when, he says, one infers the presence of a philosopher from a beard).

Let us now discuss Hoffmann's *Vernunftlehre* (1737). Hoffmann is mathematically more sophisticated than Müller and he is more sensitive to the difference between arithmetical, algebraic, and geometrical inferences.[93] He distinguishes twelve main types of inference.[94] One of these types is syllogistic inference[95] and another one is mathematical inference.[96] The bottom line on syllogistic inference in mathematics is similar to that of Rüdiger and Müller (same opposition between existential and causal inferences, etc.) but it is more relaxed. Indeed, Hoffmann grants that at times one needs to use syllogisms in mathematics or mathematical reasonings in philosophy.[97] Hoffmann puts less emphasis than Rüdiger or Müller on the issue of syllogisms in mathematics. Indeed, in the long discussion of the difference between mathematics and philosophy that opens his *Vernunftlehre*, the issue is not emphasized.

We conclude this section with a mention of the surprising position defended by Crusius.[98] Whereas there are many elements that are in common between Rüdiger, Müller, Hoffmann, and Crusius, the latter, in his treatment of the opposition between mathematics and philosophy given in *Weg zur Gewissheit und Zuverlässigkeit der menschlichen Erkenntnis* [*A Path to the Certainty and the Reliability of Human Knowledge*] (1747), makes a rather surprising claim. Indeed, in section 10, Crusius gives a long list of features that distinguish mathematics and philosophy.[99] The ninth distinction claims the following:

> In addition to these conversions, in mathematics one makes use almost exclusively of subsumption arguments or of syllogisms properly called. We will however show at the appropriate place that the human intellect uses, and must use, additional inferences that are indispensable in philosophy.[100]

We thus see that Crusius' position runs contrary to that defended in the tradition he belongs to. Whether Crusius holds a coherent position should be studied more carefully. Indeed, the above quote stands in contrast to what

---

[93] See Hoffmann (1737, Part II, chapter 5, Sec. X, pp. 868–79 on mathematical inference).
[94] See Hoffmann (1737, Part II, chapter 5, §21, p. 753).
[95] See Hoffmann (1737, Part II, chapter 5, Sec. V, pp. 806–34).
[96] See Hoffmann (1737, Part II, chapter 5, Sec. X, pp. 868–79).
[97] See Hoffmann (1737, Part II, chapter 5, §175, p. 876).
[98] On Crusius, see also Hogan (2020).   [99] Crusius (1747: 18–21).
[100] Crusius (1747: 20).

he does when he treats of arithmetical reasoning, for on occasion he gives examples of such reasoning (adding 2, 3, and 4 to obtain 9) and claims that it would be ridiculous to consider them syllogisms.[101]

At this point, we will not continue this line of investigation on early eighteenth-century German philosophy. In the next chapter, we discuss two more German philosophers, who play an important role in our story: Lambert and Kant. Lambert is remarkable for the attempt to introduce relations in his logical calculus. From this point of view, even though his logical writings exerted a very poor influence, we may consider him as a kind of ideal link connecting the scholastic and the late scholastic logicians with Augustus De Morgan and Charles S. Peirce. As for Kant, the issue is deeply intertwined with a number of interpretative debates on Kant's philosophy of mathematics, in general, and with the specific problem of the use of syllogisms in mathematical proofs. To prepare the ground to our discussion of Kant on syllogistic proofs and mathematics, we present in Section 6.2 a sketch of Kant's relationship to traditional logic.

5.3.2.3 Appendix: Note (d) in Rüdiger's *De Sensu Veri et Falsi* (1722)
Note (d) in Rüdiger's *De Sensu Veri et Falsi* (1722) is made up of ten sections. We will translate them in order to convey the tenor of Rüdiger's argumentative style:

> (d) [§1]. Syllogistic reasoning is metaphysical and teaches only the existence of things, and since no entity can lack *existence* and *quantity*, all entities are subject to metaphysical as well as to mathematical reasoning. However, they do not require such reasoning for in both of them the consideration of numerous entities is useless and foreign since what such consideration provides is already known to ordinary knowledge as will be shown momentarily by means of some examples. On the other hand, mathematics, natural philosophy and moral philosophy teach the essences and either do not concern themselves with existence or they do so only as a second intention. Thus, syllogism can be present also among these subjects but almost without any effect on the discipline itself, for it is completely unable to extract mathematical, physical and moral essences. For this reason I have claimed that mathematical reasoning *as such* does not admit the laws of syllogism. Mathematical reasoning, in fact, excels above all in constructing *theorems* and in solving *problems* but in these there is no useless application of

---

[101] Crusius (1747: chapter 8, §342, p. 611).

syllogism just as it is lacking in proving effects [from causes] and means [towards ends]. If, on the other hand, the issue is that of applying already constructed theorems, especially in case such an application is not trivial, then this syllogistic chatter plays an important role, God willing! But this happens only infrequently.[102]

In §2, Rüdiger shows how to syllogize Pythagoras' theorem. Rüdiger discusses it at length as an example of how the sensual circumstances can only be mastered through a computation of equalities and this allows him to conclude that the example shows how mathematical ratiocination differs from the syllogistic one. The proof relies on the proof previously given by Rüdiger whose formulation in the second edition we have not given. However, the syllogized proof can be read even without having the original proof available:

§2. So that all of this might become even more evident let me present and prove Pythagoras' theorem with twenty one syllogisms.

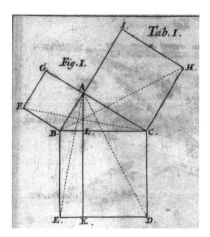

(1) All the angles of squares are right angles. The angles A.B.F. and C.B.E. are angles of squares. Thus, the angles A.B.F. and C.B.E. are right angles.

(2) All right angles are equal; the angles A.B.F. and C.B.E. are right angles; thus: the angles A.B.F. and C.B.E. are equal.

---

[102] Rüdiger (1722: 287, Book II, chapter IV, §III, note d).

(3) All aggregates that are obtained adding the same thing to two equal things are equal; the angles F.B.C. and A.B.E. are such aggregates; thus, the aforementioned angles are equal.

(4) All the sides of the same square are equal; the side F.B. of the triangle F.B.C. and the side A.B. of the triangle A.B.E. are sides of the same square; thus, the aforementioned sides are equal.

(5) Granted the major premise, the same holds of the side B.C. of the triangle F.B.C. and of the side B.E. of the triangle A.B.E.

(6) If we assume that two triangles have each two sides equal to two sides of the other and at the same time also the angles contained by the two corresponding sides equal to one another then those two triangles are equal to one another. The triangles F.B.C. and A.B.E. are two triangles having each two sides equal to two corresponding sides of the other (that is, F.B. = A.B. by syll. 4 and B.C. = B.E. by syllog. 5) and the angles contained by the two sides are also equal (that is: F.B.C. = A.B.E. by syll. 3); Thus, the triangles F.B.C. and A.B.E. are equal.

(7) Every parallelogram constructed with a triangle on the same basis and between the same parallels is double that triangle. The parallelogram F.G.B.A. is constructed with a triangle (F.B.C.) on the same basis (F.B.) and it is contained between the same parallels (G.C. and F.B.); Thus, such a parallelogram is double that triangle.

(8) From the same major premise one infers in an analogous way that the parallelogram B.E.K.L. is double the triangle A.B.E.

(9) If any two things are equal their doubles are equal; the triangles F.B.C. and A.B.E. are equal by syll. 6.; Thus: the doubles of triangles F.B.C. and A.B.E. are equal.

(10) The doubles of triangles F.B.C. e A.B.E. are equal, by syll. 9; the parallelograms F.G.B.A. and B.E.K.L. are doubles the triangles F.B.C. and A.B.E. by syll. 7 e 8; Thus, these parallelograms are equal.

With ten analogous syllogisms one shows also, with the help of triangles B.C.H. and A.C.D. that the parallelograms A.I.H.C. and L.C.K.D. are equal.

(21) One concludes then with the following syllogistic proof: B.E.K.L. + L.C.K.D. are equal to F.G.B.A. + A.I.H.C. by syll. 10. and 20. B.E.C.D. is equal to B.E.K.L. + L.C.K.D. Thus, B.E.C.D. is equal to F.G.B.A + A.I.H.C. Q.E.D.[103]

---

[103] Rüdiger (1722: 287–9, Book II, chapter IV, §III, note d).

Rüdiger makes the following remarks about the proof:

§3. This entire demonstration consists of 21 syllogisms and two theorems. If one looks attentively, however, these twenty one syllogisms make almost no contribution to the mathematical end since they do not put together any quantity but rather, as it usually happens with metaphysical reasoning and syllogisms, they only show the existence or the presence of something or of a definition, that is of the square in the case of our theorem, to which syllogisms 1, 4 and 5 refer, and of the right angle, which is related to syllogism 2; or they show the existence of a common notion in the theorem, as it happens with syllogisms 3 and 9; or they bring to light the existence of another theorem contained in ours, such as syllogisms 6, 7, and 8; or of a syllogistic conclusion in the same, as syllogism 10. Syllogism 21, however, to which the Q.E.D. refers, seems truly mathematical. However, a more careful consideration shows that it says nothing else than there are twenty syllogisms in our theorem. Its major premise, in fact, repeats what was already stated before, but in no way demonstrated, by the twenty syllogisms. In reality, what needed to be demonstrated is demonstrated mathematically by the two theorems that constitute the major premises of syllogisms 6 and 7. But no one will be able to prove these two theorems syllogistically.[104]

However, Rüdiger does not ground his claim that the major premises of syllogisms 6 and 7 cannot be proved syllogistically. He does however provide general a priori considerations for his claims.

§4. True mathematical reasoning develops by means of theorems and problems while the one that concerns *applications* and takes place through *subordination* is metaphysical and is common to all the disciplines, thus including mathematics. To it one must oppose mathematical reasoning proper (which, with some differences, is also present in *physics* and *moral theory*) that must never proceed syllogistically. Lest the impression be given that I carelessly propound the claim, I will show it as follows. Everything that mathematics proves is obtained through *enumeration* and by means of the four arithmetical species: *addition, subtraction, multiplication* and *division*. By contrast, everything that is obtained affirmatively through syllogisms is obtained by means of *predicables* while all that is obtained negatively is obtained by means of *opposites*. But the numbers occurring in addition, subtraction, multiplication and division are neither predicables nor opposites.

---

[104] Rüdiger (1722: 289, Book II, chapter IV, §III, note d).

Thus, everything that has been obtained through numbers is not obtained through syllogisms. That a number is a predicable or an opposite would not be claimed even by a naïve person completely ignorant of all such arguments. Indeed, the number three quarters is not a *genus*, a *species*, a *difference*, an *accident* or a *proprium* and even if one were intent on reconducing the numbers to the opposites or even to the different, even if we granted this, what we want to prove would still be evident: a syllogistic reasoning that moves from opposites can only lead to negative propositions. But in mathematics from opposites one almost always infers positive propositions. Thus, the mathematical way of reasoning cannot in any way be syllogistic even though syllogisms [...] insinuate themselves in all disciplines and thus also in mathematics.[105]

Having given an a priori proof of his claim, Rüdiger now offers an *a posteriori* proof:

§5. If those who object to my argument were to investigate the source of syllogism, they would, I am persuaded, give up their obstination in contradicting me. But they can, however, be persuaded directly of the same result by an argument *a posteriori*. The following reasoning, in fact, is certainly mathematical: the whole circle contains 360 degrees thus half of it will contain 180 degrees. Let then, please, all the builders of syllogistic labyrinths [Daedali syllogistici] step forward in order to measure their forces in transforming the argument into a syllogism whose major or minor premise be the following premise: *the whole circle contains 360 degrees* and the followings sentence the conclusion: *half of the circle contains 180 degrees*. They will have to be careful, however, not to fall prey to what has happened sometimes to others who have contrasted me on this argument, who ended up producing an aborted syllogism with three premises and five terms, which will easily happen if they argue as follows:

1.   2.

360 degrees make a whole circle;

3.   4.

180 degrees are half of 360 degrees;

                    5.

Thus: 180 degrees are half of the circle.

---

[105] Rüdiger (1722: 289–90, Book II, chapter IV, §III, note d).

This species of syllogism, as it is clear, has five terms and thus it is not a syllogism but a monstruous syllogism. And yet it is a genuine reasoning although it is not a syllogism. Against such a clear demonstration one should not listen to either recent or ancient authors [...][106]

We skip the beginning of section 6 where Rüdiger refers in an aside to some sections of his book that we have not discussed. The discussion of the example continues in §7 where the argument is also supported by the following considerations:

§7. That such reasoning is not a syllogism appears without doubt from the fact that it cannot have a particular conclusion, (1) in a proper sense, not deriving from *subalternation* and that it be (2) *mathematical*, that is such as to insert some quantity, not the mere existence of a quantity, as it is evident from §6. Indeed, if numbers do not admit of quantity they also do not admit particularity. You will however object: try to deny that there might be a syllogism such as the following:

Every circle has 360 degrees;
Some figure is a circle;
Thus, some figure has 360 degrees.

I reply that this is not the syllogism about which I was speaking: its conclusion, in fact, is not mathematical but metaphysical and it only asserts the existence of a figure of 360 degrees. But you go on to propose yet another syllogism:

Every circle has 360 degrees;
Some two contiguous angles make a circle;
Thus, some two contiguous angles make 360 degrees.

I reply again that not even this is the syllogism about which I was speaking. Indeed, either it is absurd to say 'some two angles' or with the same right one can say 'every two angles'; and thus the conclusion is not particular in a proper sense but it is obtained by subalternation. And I fear not to state that in mathematics there cannot be a particular conclusion in the proper sense that cannot be also stated, salva veritate, also universally. Thus, mathematical reasoning rejects syllogistic rules, especially if one considers what I showed in §6, namely that concerning numbers it is not possible to conceive quantity considered distributively while syllogism is inconceivable without this quantity.[107]

---

[106] Rüdiger (1722: 290, Book II, chapter IV, §III, note d).
[107] Rüdiger (1722: 292, Book II, chapter IV, §III, note d).

Let us conclude with the last three sections of note (d):

§8. If you want a further a priori proof, it is evident that such a reasoning is not a syllogism. The form of such a reasoning, in fact, is an enumeration while the form of syllogism is *subsumption*. But subsumption is only given of predicables. But the numbers are neither predicables nor opposites. See the proof given above in chap. III, note A §.3.[108]

§9. However, in order to sincerely defend the honor of syllogism, I will also try the following. I will ask myself if in fact mathematical reasoning does not admit the rules of syllogism in the case in which the greater number is assumed in place of the genus, the lesser number in place of the species and equality for the *proprium*. Let us consider this.

Who will have something to object, in fact, to this syllogism? Six are less than ten; four are less than six; thus, four are less than ten. But if we now conclude as follows on the basis of other syllogistic laws: seven are more than six; ten are more than six; thus, seven are more than ten, it will become clear that the former argument is not a syllogism but an *enumeration*. If, in fact, this conclusion had been reached by means of a syllogism then the conclusion of the second argument should also have to be true. Furthermore, consider that the syllogism cannot conclude in any way in the second figure for the simple reason that nothing is opposite to numbers. All entities, in fact admit number while the ground on the basis of which this [the second] figure concludes concerns also the laws of syllogism. From all of this results that mathematical arguments do not admit the laws of syllogism.[109]

§10. But it might be the case that they admit laws such as: *if the greater number is taken as the species, the lesser number for the genus.* Let's try. *Ten is more than six; eight is less than ten; thus, eight is less than six.* Since this argument, composed according to the laws of syllogism, yields a false conclusion, and since there is no other way to match the numbers with the predicables, in addition to the two attempts mentioned above, it is evident that *there is no criterion* for matching mathematical reasonings with syllogistic laws.[110]

This completes the long note (d).

---

[108] Rüdiger (1722: 293, Book II, chapter IV, §III, note d).
[109] Rüdiger (1722: 293, Book II, chapter IV, §III, note d).
[110] Rüdiger (1722: 293, Book II, chapter IV, §III, note d).

6

Lambert and Kant

6.1 Johann Heinrich Lambert (1728–1777) and the Treatment of Relations in His Logical Calculus

In his account of the origins of algebraic logic in the eighteenth century, Theodore Hailperin remarks:

> As with Leibniz, the logical vision of his 18th century followers hardly extended beyond the subject predicate form of sentences and the syllogistic doctrine. Yet even in this circumscribed ambit their results fell short of logical clarity and deductive rigor displayed in the Leibniz manuscripts. Lambert, referred to by one of his admirers as "zweyter Leibniz" came the closest.[1]

Since Leibniz, indeed, it was Johann Heinrich Lambert, who attempted to construct a logical calculus pursuing further the Leibnizian project for a universal characteristic. Even though Lambert did not have any acquaintance with Leibniz's essays on logic that were (partially) edited for the first time in 1903, he proposed a calculus that recalls some features of Leibniz's logical calculi. A reliable account of Lambert's calculus has been made by Clarence I. Lewis and, more recently, by Theodore Hailperin himself; therefore, we do not need to dwell on it.[2] We limit ourselves to say something about Lambert's treatment of relations and relational sentences.

Concerning the ontological nature of relations, Lambert agrees with Leibniz:

> The concept of a relation refers mainly to a thinking entity and always presupposes at least two things that are compared the one with the other. These two things are not themselves the relation, and the relation is neither in the one nor in the other of them, but it is simultaneously *between* them;

---

[1] Hailperin (2004: 339).   [2] Cf., Lewis (1960: 18–29); Hailperin (2004: 338–42).

Syllogistic Logic and Mathematical Proof. Paolo Mancosu and Massimo Mugnai, Oxford University Press.
© Paolo Mancosu and Massimo Mugnai 2023. DOI: 10.1093/oso/9780198876922.003.0007

and by itself it is only something ideal, independently of the fact that what is in the things as a foundation could be real.[3]

This passage contains bits of traditional scholastic doctrine about relations (the use of the notion of 'foundation,' for instance), and echoes an analogous passage of Leibniz's fifth letter to Clarke, where Leibniz says that a (binary) relation cannot 'inhere' in both the related subjects.[4] For Lambert too, as for Leibniz, relations are "something ideal" and always presuppose a thinking subject.

As Leibniz before him, Lambert states that if we want to ensure the greatest generality to the algebraic calculus, the letters do not have to be exclusively interpreted as numbers or quantities, they may also express "things and relations" and *operations* on them.[5]

Lambert interprets sentences of the general form 'A is B' according to the *intensional* reading (in Leibniz's sense): 'the attribute denoted by B inheres in the subject denoted by A,' and employs the symbol '>' to denote the copula. Thus, 'A is B' becomes:

A > B.

Because in an algebraic calculus it is easier to work with the identity relation than with inherence, Lambert resorts to the same solution adopted by Leibniz (and later by Boole[6]) of transforming inherences into identities:

Suppose given, for example, the proposition
> *T is f,*
>
> whose subject, besides the property *f*, is composed of other properties [...]
> To reduce it to an identity, we need to add to the predicate *f* a letter *t*
> containing all the remaining properties of *T*; then, we will have the identity
> $T = ft.$[7]

Consequently, the proposition 'some A are B' will be expressed as: '$mA = nB$,' in the same way as Leibniz represents the particular affirmative proposition.[8]

Introducing relations and relational sentences, Lambert adopts the following symbolism: he uses letters of the Greek alphabet to denote relations and

---

[3] Lambert (1965, IV, 2: 41).   [4] Leibniz (1965, 7: 401).   [5] Lambert (1765: 441–4).
[6] Cf., Boole (1847: 21).   [7] Lambert (1765: 454).
[8] See, for example, Leibniz (1999, 4A: 782).

letters of the Latin alphabet to denote concepts; whereas the symbol '::', interposed between a Greek and a Latin letter, expresses the subsisting of the relation. Thus, supposing that i = 'fire'; h = 'heat'; α = 'cause,' Lambert expresses as follows the proposition that fire causes heat:

$$i = \alpha :: h.$$

With the symbol '::' Lambert represents the application of a relation to a concept. Accordingly, he proposes that the converse ("the opposite") of a relation should be designated putting the symbol of the relation under the letter denoting the concept, separated by a straight line:[9]

$$\frac{\alpha :: h}{\alpha}$$

In a chapter devoted to the investigation of 'the nature of metaphysical relations,' he introduces even the powers of a relation:[10]

Suppose φ = the letter representing a given relation. Then we have, for example

$a = \varphi :: b.$

I. Suppose now $b = \varphi :: c$; then, we have

$a = \varphi :: \varphi :: c.$

Suppose furthermore $c = \varphi :: d$; then we have:

$a = \varphi :: \varphi :: \varphi :: d$

and so on.

Let us write, for the sake of brevity:

$\varphi :: \varphi = \varphi^2$ and $\varphi :: \varphi :: \varphi = \varphi^3$, etc.,

to show the degree according to which the concepts a, b and c are related.[11]

As Theodore Hailperin has remarked, however, besides introducing the converse and the powers of a relation Lambert does not mention any other operation or combination of relations (not even '$\varphi :: \psi :: b$,' "corresponding to what is now called the relative product of φ and ψ").[12] Thus, Lambert's attempt to develop a logic of relations, if compared with the results obtained by Lambert's predecessors in the same field, does not evolve besides an embryonic, even though very interesting and promising, state.

---

[9] Lambert (1782, vol. I: 19).   [10] Lewis (1960: 28–9).   [11] Lambert (1782, vol. I: 27).
[12] Hailperin (2004: 341).

6.2 Kant and Traditional Logic

We now take what, at first glance, may appear like a detour: our next target will be Immanuel Kant's views about logic and geometrical demonstration. Even though we do not know of any explicit attempt on Kant's part to prove syllogistically a Euclidean theorem or to disprove such a possibility, in the second half of the twentieth century the Kantian views on logic and the nature of mathematics gave rise to a debate that has interesting points of contact with the main topic of our story.

According to Michael Friedman, one of the sharpest interpreters of Kant's philosophy, Kant, at a certain point of his philosophical development, became aware of the inadequacy of traditional logic for performing mathematical (geometrical) demonstrations, and this very fact compelled him to introduce the notion of a priori intuition and a priori synthetic judgments.[13] If this were the case, Kant should have a preeminent place in our story, being the first to recognize that the syllogism as well as the entire traditional (monadic) logic do not possess the resources necessary to prove mathematical theorems. Friedman's thesis is intriguing and as a preliminary to a discussion of it, we proceed by first giving a general account of Kant's views about formal logic. We will then focus on some relevant interpretations concerning the nature of mathematical demonstrations in Kant. Finally, we will conclude with a discussion of Friedman's thesis and some recent work on the foundations of geometry in so far as it bears on the thesis.

The task of describing Kant's attitude towards formal logic is made easier by the fact that, on this issue, Kant did not dramatically change his views in passing from the pre-critical to the critical period. Thus, to reconstruct Kant's ideas on formal logic we may rely quite confidently not only on what emerges on this topic from the *Critique of Pure Reason* (1781) but even from pre-critical writings such as, for instance, the *False Subtlety of the Four Syllogistic Figures* (1762) or some student notebooks which report some of Kant's university lectures on logic.

Kant taught logic at the University of Königsberg for about thirty years, but he never published or wrote an essay entirely devoted to this discipline.[14] For his courses he employed Georg Friedrich Meier's *Auszug aus der*

---

[13] In the sections on Kant we have to presuppose a number of key notions of Kant's philosophy. This is inevitable, for an attempt to introduce all the notions required for our treatment would require at least an additional chapter.

[14] With the exception of *The False Subtlety of the Four Syllogistic Figures* (1762; see Kant (1992b: 84–105)).

Vernunftlehre [*Excerpt from the Doctrine of Reason*] and charged his assistant Gottlob Benjamin Jäsche with the task of composing a text based on his classroom lectures (later known as *Jäsche Logic*). Kant filled Meier's book with marginal remarks and inserted between its pages several sheets of paper containing notes on topics related to logic. These materials are enriched by some surviving student notebooks from Kant's lectures and by scattered observations on logic included in the *Critique of Pure Reason* and other major works. Thus, if one aims to reconstruct Kant's views on logic, one is forced to take into account a bunch of heterogeneous texts whose degree of reliability (as in the case of the lecture notes taken by his students, for instance) is not so high.

Therefore, the first question that we meet when considering Kant's attitude toward logic is why, notwithstanding the fact that he taught logic for so many years, he did not compose an essay devoted to this discipline. A plausible answer is that for Kant logic[15] did not occupy such a prominent place among the philosophical disciplines as it did, for instance, in the case of Leibniz.

According to Kant, logic "shows the rules for the use of the understanding and reason, which can themselves be cognized a priori and without experience, since they do not depend on it."[16] Logic does not contain the rules in accordance with which we actually think but the rules according to which we ought to think: it is a *normative* discipline and cannot derive its principles from experience (hence not from psychology).[17] As Kant emphasizes on several occasions, logic is a *formal* discipline and it mainly concerns itself with the *form of thought*.

Even though on this point Kant clearly distances himself from Wolff, at first glance he seems to have quite a traditional view of logic. To describe the basic structure of a proposition or *judgment*, he employs the categories of *subject* and *predicate*:

> To compare something as a characteristic mark with a thing is *to judge*. The thing itself is the subject; the characteristic mark the predicate. The comparison is expressed by means of the copula *is* or *are*. When used absolutely, the copula designates the predicate as a mark of the subject. If, however, it is combined with the sign for negation, the copula then signifies that the

---

[15] We are, of course, restricting our discussion to formal logic and eliminating from consideration transcendental logic.

[16] Kant (1992a: 12) and see: "Logic is called a science because its rules can be proved by themselves, apart from all use[,] *a priori*."

[17] Kant (1992a: 13).

predicate is a characteristic mark which is incompatible with the subject. In the former case, the judgement is affirmative, whereas in the latter case the judgement is negative.[18]

Whereas a *judgment* consists in attributing a *characteristic mark* (*Merkmale*) to a *thing* (a *subject*), a *syllogism* is a judgment made by means of a *mediate* characteristic mark:

> *Every judgement which is made by means of a mediate characteristic mark is a syllogism.* In other words, a syllogism is the comparison of a characteristic mark with a thing by means of an intermediate characteristic mark. This intermediate characteristic mark (*nota intermedia*) in a syllogism is also normally called the *middle term* (*terminus medius*) [...][19]

In other words, a syllogism is an inference composed of three judgments (three categorical propositions), in the usual way. According to Kant, a syllogism obeys two different principles, depending on whether it is either affirmative or negative:[20]

> [...] the first general rule of all affirmative syllogisms is this: *A characteristic mark of a characteristic mark is a characteristic mark of the thing itself* (*nota notae est etiam nota rei ipsius*). And the first general rule of all negative syllogisms is this: *that which contradicts the characteristic mark of a thing, contradicts the thing itself* (*repugnans notae repugnat rei ipsi*).[21]

Given a sentence like, for instance, 'Socrates is a man,' the name *Socrates* refers to the 'thing' called 'Socrates' and *man* refers to a characteristic mark of this thing. Since 'mortal,' in its turn, is a characteristic mark that composes the characteristic mark corresponding to 'man,' it follows that the 'thing' called *Socrates* has the characteristic mark of being mortal. The rule for negative syllogisms is analogous and can be easily derived from the affirmative one.

Kant considers these two principles as more fundamental than the traditional *dictum de omni et nullo* and therefore, as he writes, they 'precede' the *dictum*. In the following passage, Kant resumes these principles and gives his own interpretation of the *dictum*:

---

[18] Kant (1992b: 89).   [19] Kant (1992b: 90).
[20] In this case, one of the two premises and the conclusion are negative.   [21] Kant (1992b: 91).

The basic rules of all affirmative inferences are these: 1. *Nota notae rei est nota rei ipsius*[.] The only basic rule, however, 2. of all *negative* inferences will thus be: *Repugnans nota notae rei, repugnat rei ipsi*[.] This is now the basic rule of all inferences of reason. Logicians have a certain *dictum de omni et nullo*, which they take to be the very first ground of all *ratiocinia*. But these basic rules that have been presented precede even this. Among logicians, the *dictum de omni et nullo* runs thus: what belongs to a universal concept belongs also to all concepts that are contained under it.[22]

Therefore, for Kant, a *ratiocinium*, or syllogism, is composed of three steps: first, an intermediate mark is compared with a remote mark, then the thing (the subject) is compared with the same intermediate mark and, finally, the remote mark is compared with the thing:

> A *judicium*, namely, involves an *immediate* clear mark, but a *ratiocinium* involves a *mediate clear mark*. Consequently, a *ratiocinium* is the relation of the *nota remota* by means of a *nota mediata, vel intermediata* to a certain given concept of a thing. In a *ratiocinium*, then, these 3 actions occur: 1st the *nota proxima* or *intermedia* is compared with the *nota remota*, 2nd the *nota proxima* is compared with the thing, et 3rd the *nota remota* is compared with the thing itself.[23]

This is only an exotic way of presenting the first figure syllogisms of the traditional syllogistic:

M (= intermediate mark) P (= remote mark)
S (= subject, the 'thing') M (= intermediate mark)
S (= subject, the 'thing') P (= remote mark).

Since the first figure syllogism perfectly fits the above principles concerning the characteristic mark of a characteristic mark, and since the remaining figures can be proven valid by means of the first one (with the aid of some immediate inferences), Kant considers superfluous and, up to a certain point, *false* the syllogistic figures different from the first. As he writes in the *False Subtlety*, insofar as these figures are "syllogistic rules in general," they are "correct," but when "construed as containing a simple and pure inference, they are mistaken."[24] Kant motivates this claim remarking that the figures different

---

[22] Kant (1992a: 229).   [23] Kant (1992a: 228).   [24] Kant (1992b: 99–100).

from the first overtly violate what he considers the logic's "distinctive purpose," namely "that of reducing everything to the simplest mode of cognition":

> The purpose of logic, however, is not to confuse but to clarify; its aim is not
> to obscure but clearly to reveal. Hence, these four modes of inference ought
> to be simple, unmixed and free from concealed supplementary inferences. If
> they do not satisfy these conditions they are not to be granted [...][25]

Besides the reasons adduced, however, Kant's claim about the primacy of the
first syllogistic figure is not particularly original, and, in general, his views on
the syllogistic machinery are far from new. He distances himself, however,
from the tradition at least on two points:

(1) Even though Kant considers as basic categorical propositions in subject-
predicate form, because they "constitute the matter of the remaining judgments," he does not believe "that both hypothetical and disjunctive judgments
are nothing more than various clothings of categoricals and hence may be
wholly traced back to these latter." In other words, for Kant, categorical,
hypothetical and disjunctive judgments "rest on essentially different logical
functions of the understanding and must therefore be considered according to
their specific difference."[26]

(2) He does not attribute great importance to the traditional four syllogistic
figures.

Point (1) is clearly witnessed by the following passages:

> All relations of thinking in judgments are those *a*) of the predicate to the
> subject, *b*) of the ground to the consequence, and *c*) between the cognition
> that is to be divided and all of the members of the division. In the first kind of
> judgment only two concepts are considered to be in relation to each other, in
> the second, two judgments, and in the third, several judgments. The hypo-
> thetical proposition "If there is perfect justice, then obstinate evil will be
> punished" really contains the relation of two propositions, "There is a perfect
> justice" and "Obstinate evil is punished." Whether both of these propositions
> in themselves are true remains unsettled here. It is only the implication that
> is thought by means of this judgment. Finally, the disjunctive judgment
> contains the relations of two or more propositions to one another, though
> not the relation of sequence, but rather that of logical opposition, insofar as
> the sphere of one judgment excludes that of the other, yet at the same time

---

[25] Kant (1992b: 99).   [26] Kant (1992a: 601).

the relation of community, insofar as the judgments together exhaust the sphere of cognition proper [...][27]

The hypothetical judgment is composed of two problematic ones; the disjunctive judgment, however, is composed of two or more judgments. In the hypothetical judgment I consider the combination of two judgments as ground and consequence. In the disjunctive judgment all the categorical judgments are members of the division [;] it is to be noted, however, that in the case of hypothetical and disjunctive judgments, the judgments cannot be transformed into categorical judgments again. - The matter of a categorical judgment consists of two concepts, the form in the relation in which the one concerns the subject, the other the predicate. E.g., All men are mortal. - In the hypothetical judgment the matter consists of two judgments. E.g., If the soul is corporeal, then there is no hope of the necessity of another life.[28]

Kant is aware that some authors think that it is possible to transform hypothetical into categorical propositions, but when this transformation is done, "what is maintained is no longer the same."[29] We cannot substitute, for instance, the hypothetical proposition 'If something is a man, then it is mortal' with the categorical 'All men are mortal,' because in the first proposition "it is problematic whether something is mortal. Being mortal is not maintained categorically but holds only when being man holds." Therefore,

it is completely different with categorical propositions than with hypothetical ones. In a hypothetical proposition it is not maintained at all that something is, but that it is if something, namely, the ground, is accepted. In the case of categorical propositions, however, there is no settled condition. They are judgments essentially distinct from one another, then.[30]

Point (2) is made clear by this passage, from the *Critique of Pure Reason*:

The widespread doctrine of the four syllogistic figures concerns only the categorical inferences, and, although it is nothing more than an art for surreptitiously producing the illusion of more kinds of inference than that in the first figure by hiding immediate inferences (*consequentiae immediatae*) among the premises of a pure syllogism, still it would not have achieved any special success by this alone if it had not succeeded in focusing attention

---

[27] Kant (1998: 208).   [28] Kant (1992a: 373).   [29] Kant (1992a: 374).
[30] Kant (1992a: 374).

exclusively on categorical judgments as those to which all others have to be related, which according to §9, however, is false.[31]

Thus, Kant does not regard the traditional syllogism of Aristotelian origin as the cornerstone of every logical proof, nor does he consider the categorical judgments as the only basic judgments to which all others are reducible. From this point of view, Kant distances himself from the logical orthodoxy represented by Wolff. His attitude toward the logical tradition, however, as we will see soon, is not so firm as it may appear from the above passage of the *Critique of Pure Reason*.

The acceptance of the three kinds of basic judgments—categorical, hypothetical, and disjunctive—raises a question that emerged in the discussions of some recent interpreters of Kantian philosophy. To see what the nature of this question is, let us first recall Kant's famous distinction between *analytic* and *synthetic* judgments.

In the *Introduction* to the *Critique of Pure Reason*, we read:

In all judgments in which the relation of a subject to the predicate is thought (if I consider only affirmative judgments, since the application to negative ones is easy), this relation is possible in two different ways. Either the predicate B belongs to the subject A as something that is (covertly) contained in this concept A; or B lies entirely outside the concept A, though to be sure it stands in connection with it. In the first case I call the judgment analytic, in the second synthetic. Analytic judgments (affirmative ones) are thus those in which the connection of the predicate is thought through identity, but those in which this connection is thought without identity are to be called synthetic judgments. One could also call the former judgments of clarification and the latter judgments of amplification, since through the predicate the former do not add anything to the concept of the subject, but only break it up by means of analysis into its component concepts, which were already thought in it (though confusedly); while the latter, on the contrary, add to the concept of the subject a predicate that was not thought in it at all, and could not have been extracted from it through any analysis [...][32]

Here, Kant explicitly restricts his attention to *true, affirmative* judgments "in which the relation of a subject to the predicate is thought," that is, to

---

[31] Kant (1998: 251); see even Kant (1992a: 222): "This whole designation is mere pedantry (Barbara[,] Celarent[, etc.]), and what is more, all of syllogistic is mere pedantry."

[32] Kant (1998: 130).

judgments that correspond to true, affirmative categorical propositions of the syllogistic tradition. Thus, the question poses itself whether the distinction between *analytic* and *synthetic* judgments applies to hypothetical and disjunctive judgments, as well.[33] And this also leads to the more general question of whether logic, according to Kant, is the exclusive realm of analytic judgments or whether it includes even the synthetic ones.

As we have seen, Kant, even though stating that hypothetical and disjunctive judgments are not reducible to the categorical ones, still continues to consider these latter as basic:

> Categorical judgments constitute the *basis* of all the remaining ones. Here the relation of subject with predicate is indicated.[34]

One of the reasons why Kant considers the categorical judgments as "the basis" of the non-categorical ones is that, according to the scholastic tradition, the elementary form of any proposition has to be in the form of 'subject—(copula)—predicate.' Hence, from this point of view, hypothetical and disjunctive judgments are both composed of categorical judgments. That the disjunctive judgments are composed of categorical judgments, indeed, is explicitly asserted in one of the transcripts of Kant's logical lectures, which we have already quoted:

> The hypothetical judgment is composed of two problematic ones; the disjunctive judgment, however, is composed of two or more judgments. In the hypothetical judgment I consider the combination of two judgments as ground and consequence. In the disjunctive judgment all the categorical judgments are members of the division[;] it is to be noted, however, that in the case of hypothetical and disjunctive judgments, the judgments cannot be transformed into categorical judgments again.[35]

As we may see, in the final remark of this text, Kant stresses that hypothetical *and* disjunctive judgments "cannot be transformed into categorical judgments *again*" (emphasis added): this means that hypothetical and disjunctive judgments cannot be reverted to the categorical form, *from which they originate*.

---

[33] The issue of the application of the analytic/synthetic distinction is discussed in Proops (2005); see also Anderson (2005, 2015) and Longuenesse (1998).
[34] Kant (1992a: 373).   [35] Kant (1992a: 373).

Thus, it is confirmed that, according to Kant, hypothetical and disjunctive judgments are based on categorical judgments.

The distinction between *analytic* and *synthetic* judgments, however, seems to be unsuitable for hypothetical and disjunctive judgments. Hypothetical judgments are subordinated to the relation of ground to consequence, whereas disjunctive judgments depend on the relation of *opposition* and are instances of *logical division*:

> Those judgments where one thinks the relation of two judgments with one another are called, in fact, judgments of relation (*relative*)[;] now these judgments consider the relation of one judgment to the other, either as to connection, and then they are *hypothetical* judgments, or as to *opposition*, and then they are *disjunctive judgments*. In the case of the former, one always finds the relation of ground to consequences. Now in conditioned judgments, that which contains the ground is called *antecedens* or also *prius*. That which contains the consequences, however, is called in these judgments *consequens* or *posterius*. As for what concerns disjunctive judgments, in them the relation of opposition is again of two kinds[;] either it is a relation of two, or of more judgments, which contradict one another[;] but we will show that a true disjunctive judgment occurs only with two and not with several judgments, for a true and pure contradiction cannot occur except with two concepts that stand exactly in opposition to each other. [...] A disjunctive judgment is nothing more than a representation of a logical division, or *representatio divisionis logicae*.[36]

According to the *Blomberg Logic*, Kant calls 'ordinary reasoning' any inference that employs categorical judgments and *extra-ordinary reasoning*, namely hypothetical and disjunctive inferences:

> All *ratiocinia extraordinaria* are either *hypothetica* or *disjunctiva*. For in them either the relation of connection or the relation of opposition is indicated; if it is the first, then it is called a *hypothetical* inference of reason, but in the other case a *disjunctive* one. Now the *judicium hypotheticum* consists only of 2 propositions and hence no *judicium hypotheticum* can produce a *ratiocinium*. In the *judicium hypotheticum* only the *consequentia* needs to be proved, for in this judgment nothing is enunciated or maintained. But with *judicia categorica* not only the *consequentia* but also the

---

[36] Kant (1992a: 222–3).

assertion itself must be proved from correct grounds. The conditioned inference is not an inference of reason but instead only the *substratum* for the inference of reason, for it must first be proved *per ratiocinium*.[37]

Here, Kant clearly claims that inferences based on categorical judgments have a kind of prevalence over inferences based on hypothetical and disjunctive judgments. As the editors of Kant's logical papers point out, Kant's criticism of the tradition is developed only imperfectly: he fails, for instance, in formulating a principle that will cover hypothetical, disjunctive, and categorical syllogisms; and "contrary to his own insistence, he often speaks as though there were only categorical judgments and categorical syllogisms."[38]

6.3 Kant on Syllogistic Proofs and Mathematics

To raise the question of whether Kant thought that mathematical proofs could be carried out syllogistically is to enter one of the most subtle and debated areas of Kantian scholarship. We do so not with the hope of resolving the issue but merely to frame it in a careful manner so that even a reader who is not a Kantian specialist can understand what is at stake. Our main interest is to articulate what the historical Kant might, rightly or wrongly, have thought about the issue of the syllogistic representability of mathematical proofs. This explains why, when discussing contemporary interpretations of Kant's philosophy of mathematics, we will always try to keep clear the distinction between Kant's original theory and contemporary interpretations motivated by preserving the "essence" of Kant's theory in the face of challenging new results emerging from nineteenth- and twentieth-century developments in mathematics and logic. In other words, we use the contemporary discussion as a guide to a probing analysis of the historical Kant (where we obviously remain mindful of the need to avoid anachronisms).

Kant's philosophy of mathematics can be summarized as comprising two main theses. The first is that mathematical judgments are synthetic a priori. The second, related thesis, is that mathematics proceeds by construction of concepts in intuition whereas philosophy is rational knowledge based on concepts.[39] There is agreement among scholars about these two theses but

---

[37] Kant (1992a: 230).   [38] Kant (1998: 15).

[39] Cf., Kant (1998: 630; A713/B741): "**Philosophical** cognition is **rational cognition** from **concepts**, mathematical cognition that from the **construction** of concepts. But to **construct** a concept means to exhibit *a priori* the intuition corresponding to it. For the construction of a concept, therefore, a

strong divergences as soon as one tries to interpret what is meant by them. Let us try to exemplify these divergences taking syllogization of mathematical reasoning as our starting point. In order to avoid the risk of framing the issue in anachronistic terms let us ask what Kant's position would have been concerning some of the burning issues in this area that we have discussed. We have seen that for many philosophers and mathematicians, one question was whether geometrical theorems found in Euclid could be recast in syllogistic form. Here we must however emphasize that, in agreement with Proclus' distinction of the parts occurring in Euclidean theorems and problems, the majority of contributions in this area were focused on whether the *apodeixis*, that is, the last step in the Euclidean proof, could be recast in syllogistic form. Thus, a first question for the Kantian scholar is:

(Q1)   Would Kant accept or deny that the *apodeixis* in Euclidean propositions can be carried out merely by syllogistic means?

While restricting the focus to the *apodeixis* makes perfect sense when referring to several of the authors we have discussed, Kant was also aware of more radical claims on behalf of syllogistic logic in mathematics and beyond. In particular, Wolff had not only claimed that all of mathematics could be syllogized (i.e., not just the *apodeixis* in Euclidean proofs) but that even the process of finding the truth of a mathematical proposition followed a syllogistic pattern (i.e., was controlled by a chain of syllogisms). Thus, the second question for the Kantian scholar is:

(Q2)   Would Kant accept or deny that mathematical reasoning, now taken to encompass any mathematical presentation offered in mathematical practice (and Kant mainly discussed the Euclidean practice in geometry), can be accounted for appealing to syllogistic arguments?

We must, of course, immediately flag that a satisfactory reply to the above questions will also have to address what exactly Kant would include in syllogistic logic (something which motivated our long excursus on Kant's conception of logic and syllogistic reasoning), its relation to formal logic, and the relation between the concept of analyticity and pure logic. The only

non-empirical intuition is required, which consequently, as intuition, is an **individual** object, but that must nevertheless, as the construction of a concept (of a general representation), express in the representation universal validity for all possible intuitions that belong under the same concept."

theses that we will take for granted are that all of syllogistic logic, *apud Kant*, is included in formal logic and that, according to Kant, all analytic truths are logical (we express skepticism about the converse at the end of this section).

Let us begin with the second question (Q2) to illustrate why the issue is complex. Starting from the early twentieth century two major interpretative traditions have opposed each other on the issue of the role of pure intuition in Kant's philosophy of mathematics. The first tradition goes back to Russell and Couturat (1904, 1905). The main core of this interpretative line is that Kant finds himself forced to appeal to a priori intuition in order to compensate[40] for the inherent limitations of the subject-predicate analysis of statements and of the logic of his time. In section 433 of *Principles*, Russell writes:

> Broadly speaking, the way in which Kant seeks to deduce his theory of space from mathematics (especially in the *Prolegomena*) is as follows. Starting from the question: "How is pure mathematics possible?" Kant first points out that all the propositions of mathematics are synthetic. He infers hence that these propositions cannot, as Leibniz had hoped, be proved by means of a logical calculus; on the contrary, they require, he says, certain synthetic *a priori* propositions, which may be called axioms, and even then (it would seem) the reasoning employed in deductions from the axioms is different from that of pure logic.[41]

The Russellian characterization, however debatable in some of its details, has the advantage of articulating clearly one of the most striking aspects of this interpretation of Kant, namely that a priori intuition is not only needed at the level of the axioms but rather takes a role in the proof itself. Russell emphasizes this point a little later:

> What is essential, from the logical point of view, is, that the a priori intuition supply methods of reasoning and inference which formal logic does not admit; and these methods, we are told, make the figure (which may of course be merely imagined) essential to all geometrical proofs.[42]

Among the causes Russell adduced for diagnosing Kant's being led astray was the sad state of logic of his time:

---

[40] We borrow the use of "compensation" from Hogan (2020), which contains in the first section an excellent overview of the traditions we are discussing.
[41] Russell (1903: 456).   [42] Russell (1903: 457).

Formal logic was, in Kant's day, in a very much more backward state than at present. It was still possible to hold, as Kant did, that no great advance had been made since Aristotle, and that none, therefore, was likely to occur in the future. The syllogism still remained the one type of formally correct reasoning; and the syllogism was certainly inadequate for mathematics.[43]

Clearly, Russell identifies Kant's logic with the theory of the syllogism. The claim can be questioned but we need not worry about it just now. Russell, relying on the new logic of relations, proudly declared that his views were "on almost every point of mathematical theory, diametrically opposed to those of Kant" (456) By making the distinction between logical geometry and physical geometry he could assert that in logical geometry the starting points do not require a priori intuition and that the reasoning proceeds according to logic (an enriched logic vastly expanding syllogistic theory). In *Mysticism and Logic*, Russell wrote:

> The problem with which Kant is concerned in the Transcendental Aesthetic is primarily an epistemological problem: "How do we come to have knowledge of geometry *a priori*?" By the distinction between the logical and physical problems of geometry, the bearing and scope of this question are greatly altered. Our knowledge of pure geometry is a priori but is wholly logical. Our knowledge of pure geometry is hypothetical, and does not enable us to assert, for example, that the axiom of parallels is true in the physical world.[44]

Couturat's evaluation of Kant's philosophy of mathematics (Couturat 1904, 1905) resembled that of Russell. However, by emphasizing the pervasive role of intuition in mathematical proofs, Russell gave an interpretation of Kant's theory that was more radical than the one given by Couturat. Couturat did reproach Kant for his impoverished conception of logic and criticized his appeal to a priori intuition but he interpreted a crucial passage in Kant in a way that allowed logic to play a more autonomous role in mathematical proofs than in Russell's interpretation. We are referring to the passages in *Prolegomena* (§2c) and the *Critique of Pure Reason* (B14), which led Couturat (1905: 251) to claim that Kant dangerously conceded ground to the competing theory that mathematics is analytic. As B14 has become one of

---

[43] Russell (1903: 457).   [44] Russell (1918: 119).

the major points of disagreements among the various interpretations, let us read it:

> **Mathematical judgments are all synthetic.** This proposition seems to have escaped the notice of the analysts of human reason until now, indeed to be diametrically opposed to all their conjectures, although it is incontrovertibly certain and is very important in the sequel. For since one found that the inferences of the mathematicians all proceed in accordance with the principle of contradiction (which is required by the nature of any apodictic certainty), one was persuaded that the principles could also be cognized from the principle of contradiction, in which, however, they erred; for a synthetic proposition can of course be comprehended in accordance with the principle of contradiction, but only insofar as another synthetic proposition is presupposed from which it can be deduced, never in itself.[45]

According to Couturat, Kant's concession to the rationalists (Leibniz, Wolff, etc.) was to accept that a priori intuition was not present in the proofs themselves but rather only in the principles. If the principles, as the new "logistic" showed, could be derived analytically (i.e., by mere logic) then mathematics could be shown to be analytic.

The passage at B14 is also an important element of Cassirer's critical but generous review of Couturat's work.[46] Cassirer emphasized the conflict between Russell and Couturat by accurately pointing out that since Russell believed that logic was synthetic, a successful reduction of mathematics to logic would not show the analyticity of mathematics. Moreover, while agreeing with Couturat in his reading of B14, he pointed out that a more accurate reading of the Kantian project in its entirety showed that the syntheticity of logic and mathematics would still follow from the Kantian set up even if a reduction of mathematics to logic, as advocated by the logicists, was successfully carried out. He claimed that his reading, based on the Kantian general notion of "synthesis," found Russell in agreement:

> Moreover, let us consider the specific conceptual tools of logistics: whether one considers the relation *part* to *whole*, or the general concept of *function*, or the concepts of *identity* and *difference*, – according to Russell's own

---

[45] Kant (1998: 143–4; B 14). The same passage is repeated in the *Prolegomena* (1783) (Kant 2002: 63).

[46] Cassirer (1907).

interpretation – each of these conceptual tools constitutes a fundamental relation that is not further derivable. And their meaning does not consist merely in the fact that they satisfy the *principle of contradiction* but rather in the fact that they constitute new and peculiar *posits and modes of connection* of thought and they constructively create a new content. To bring back mathematical concepts to such grounds thus means to found them on the most general primitive *syntheses*. And to prove this principle against Couturat we need do no more than appeal to Russell himself. While Couturat constantly emphasizes the fact that Kant has missed the merely analytic meaning of mathematical principles, Russell, by contrast, sees the weakness of critical philosophy in the fact that it, while already having been the first to establish the synthetic nature of *mathematical* judgments, did not extend the same claim to *logical* judgements.[47]

Cassirer, citing Pasch, agreed that the tendency of the latest geometrical research was leading to an idea of proof that rejected appeal to intuition and he was willing to accept this as a datum that philosophy had to account for. But, according to Cassirer, the epistemological problem that Kant was after remained as pressing as ever. Without entering the details of how Cassirer motivated the actuality of the Kantian project let us focus on his interpretation of B14. Here is the key passage:

> Thus, in order to achieve a valid decision concerning the analytic or synthetic character of a proposition it is never enough to consider the connection of subject and predicate only according to its *formal* aspect. One must rather always reflect at the same time on the "transcendental" origin of that knowledge that is implicit in the concept of the subject itself. What is important is not whether or not the proof can be resolved in mere tautologies, but whether the *premises* on which it is grounded are susceptible of such an analysis or not. That "mathematical proofs proceed in accordance with the principle of contradiction"; indeed, that in this consists "the nature of every apodictic certainty" was asserted by Kant with clarity. But he does not find in this fact anything that goes against his fundamental point of view: "for a synthetic proposition can of course be comprehended in accordance with the principle of contradiction, but only insofar as another synthetic proposition is presupposed from which it can be deduced, never in itself" (B14).[48]

---

[47] Cassirer (1907: 36–7).   [48] Cassirer (1907: 39).

Hogan (2020) sees Cassirer as the fountain head of those positions that insist that the injection of the synthetic (a priori intuition) element in a mathematical proof appears at the level of the principles rather than in the inferences. But we think that this is not so clear. Cassirer's position is rather subtle: first, he is willing to accept that the logistic reduction is successful and thus that, alongside with a progressive removal of a priori intuition, the syntheticity occurs in the *intellectual* syntheses[49] required to form the concepts appearing in the principles of the (logicized) mathematical proof; and second, he also seems to agree with Russell that the logical inferences are synthetic. Hence, his interpretation of B14 is consistent with the idea that the general "syntheses" will ground both the principles and the logical steps, even if the latter are reduced to mere tautologies. In other words, Russell and Cassirer might be willing to grant that logic is sufficient to bear the inferential burden of demonstrations but this does not make the process analytic. However, there is no doubt that Russell emphasizes that according to Kant a mathematical proof must be "guided" by an a priori intuition (and, he claims, contemporary logic and mathematics have proven him wrong about that), whereas Cassirer seems to be willing to give B14 an authoritative role in claiming that for Kant the inferential structure of the mathematical proof can be logical in nature.

Hogan (2020) and Friedman (1992) see the position defended by Lewis White Beck (1955–6, 1956, 1965) as paradigmatic of those who claim that in Kant's conception of mathematics the injection of a priori intuition takes place in definitions and principles while the proof itself can proceed purely logically. Friedman (1992: 81) cites the following passage from Beck:

> The real dispute between Kant and his critics is not whether the theorems are analytic in the sense of being strictly [logically] deducible, and not whether they should be called analytic now when it is admitted that they are deducible from definitions, but whether there are any primitive propositions which are synthetic and intuitive. Kant is arguing that the axioms cannot be analytic... because they must establish a connection that can be exhibited in intuition.[50]

Friedman introduces the citation by saying "This anti-Russellian view is clearly and forcefully stated by Beck."[51] The attribution of the position to Beck himself remains an implicature, and thus subject to possible Gricean

---

[49] These are the intellectual syntheses, mentioned in a previous quote, that are needed to form the concepts of "the relation *part* to *whole*, or the general concept of *function*, or the concepts of *identity* and *difference*."

[50] See Beck (1965: 89–90), originally published as Beck (1955–6).   [51] Friedman (1992: 81).

cancelation, but not so in Hogan (2020), who explicitly attributes the view to Beck and associates him to Cassirer. When Beck (1955–6) is carefully analyzed, we see that Beck's reading of B14 (and the corresponding passage in *Prolegomena*) takes place within the dialectical debate with recent anti-Kantian criticisms (Beck mentions C. I. Lewis but Russell and Couturat are also obviously in the background). Beck, who is engaged in blocking the possible dissolution of synthetic statements into analytic ones,[52] emphasizes one strand of Kant's appeal to a priori intuition as it emerges from B14, namely the idea that a mathematical proof could consist of merely logical steps, but the premises would still have to contain synthetic a priori statements in order for a synthetic a priori conclusion to be derived. However, while Beck seems to think that this is a good Kantian move in order to block the Kant's "critics" that he himself is addressing, he does not think that Kant himself coherently subscribed to such a line. Indeed, he prefaced his discussion leading to the above quote by stating the following:

> The theorems, therefore, can be called synthetic, even though they are strictly (analytically, in modern usage) demonstrable. The famous discussion of the example "$7 + 5 = 12$," two paragraphs later, is quite independent of the grounds given in the quotation [from *Prolegomena* and the associated B14] for calling the theorems synthetic. It is, in fact, inconsistent with it. In the quotation, Kant is conceding that a theorem does follow from premises by strict logic: whatever may be the nature of the premises, the internal structure of the proof is logical. But in the discussion of "$7 + 5$" Kant is arguing that a theorem does not follow logically even from synthetic axioms, but that intuitive construction enters into the theorem itself and its proof. These two theses – that an intuitive synthetic element is present in the primitive propositions, and that an intuitive synthetic process is present in demonstration – are independent of each other.[53]

Hogan himself mentions in a footnote that "Beck holds that claims regarding arithmetic contradict B14's account of proof" thereby making Beck an objector to his own interpretative position.[54] We think that no contradiction follows if one distinguishes giving an account of the historical Kant and attempting to distil a coherent "Kantian" position, grounded in some aspects of Kant, that

---

[52] See Anderson (2015: 141–3).   [53] Beck (1965: 89).
[54] Hogan (2020: 133, note 6). Friedman (1992: 83, note 46) also remarks on Beck's pointing out the inconsistency of B14 with what Kant says about arithmetic.

preserves what is essential to Kant's philosophy of mathematics and can be used to rebut recent criticisms. Consider the following theses:

(1) an intuitive synthetic element is present in the primitive propositions;
(2) an intuitive synthetic process is present in demonstration.

Beck believes that in the opposition between Kant and his recent critics only thesis (1) is of consequence while (2) is not. But he is not claiming that only the first thesis accurately reflects Kant's point of view (Beck later also reminds us that "even propositions which Kant admits are analytic belong to mathematics only if they can be exhibited in intuition [B17; Prolegomena §2 c 2]").[55]

Be that as it may, the distinction between the two theses is quite useful. For instance, we can say that Cassirer's and Couturat's positions seem to commit them to only accepting the first thesis, while Russell, Friedman, and Hogan accept both theses.

There are in any case well-known passages in the *Critique of Pure Reason* that make it difficult to claim that the injection of a priori intuition takes place only in the definitions and axioms. Consider the following passage:

> Give a philosopher the concept of a triangle, and let him try to find out in his way how the sum of its angles might be related to a right angle. He has nothing but the concept of a figure enclosed by three straight lines, and in it the concept of equally many angles. Now he may reflect on this concept as long as he wants, yet he will never produce anything new. He can analyze and make distinct the concept of a straight line, or of an angle, or of the number three, but he will not come upon any other properties that do not already lie in these concepts. But now let the geometer take up this question. He begins at once to construct a triangle. Since he knows that two right angles together are exactly equal to all of the adjacent angles that can be drawn at one point on a straight line, he extends one side of his triangle, and obtains two adjacent angles that together are equal to two right ones. Now he divides the external one of these angles by drawing a line parallel to the opposite side of the triangle, and sees that here there arises an external adjacent angle which is equal to an internal one, etc. In such a way, **through a chain of inferences that is always guided by intuition**, he arrives at a fully illuminating and at the same time general solution of the question.[56]

---

[55] Beck (1965: 91).   [56] Kant (1998: 631–2; A 717, our emphasis).

"The chain of inferences guided by intuition" seems to speak in favor of the Russellian interpretation but we will come back to it when speaking about Friedman.

Hogan sees the challenge for Cassirer/Beck types interpretations as the following: "Cassirer/Beck defenders must maintain that he [Kant] falsely believed that his logic allows rigorous formulation of mathematical proof."[57] We fully agree but do not see this as too bitter a pill to swallow, if one is inclined to go that way. And we motivate this claim by appealing to Kant's surprising silence on some of the relevant key issues and some further considerations to be adduced below.

As we have seen, Wolff defended the idea that all mathematics can be captured syllogistically. By contrast, Jungius and other German pre-Kantian thinkers such as Rüdiger had made heavy weather of the existence of non-syllogistic reasoning, especially in mathematics. Kant never explicitly says against Wolff that the idea of syllogizing geometry is a non-starter, nor does he pursue, as also remarked by Hogan, the issue of non-syllogistic reasoning, even though he recognizes the existence of disjunctive and hypothetical inferences, besides the traditional syllogistic ones. The Kantian silence on both issues might well lead one to suspect that he believed that the syllogistic logic of his time (however he construed it) could account for any informal pattern of valid inferential reasoning,[58] or put in a less ambiguous way, that there was no pattern of informal logical reasoning that syllogistic logic could not account for (Wolff for instance, as we have seen, claimed that oblique reasonings could be recaptured in terms of ordinary syllogisms). Of course, we are not saying that there are no further elements that complicate the picture enormously, and we will soon get to them, but one should not dismiss the above possibility as unrealistic.

Before moving on to Friedman, we have to take into account Hintikka's and Brittan's interpretations. Hintikka, finding some inspiration in Beth,[59] developed a unified account of Kant's philosophy of mathematics. It is a complex and elaborate account and here we will only emphasize some of its key tenets. First of all, Hintikka proposes that the distinction between the analytic and the

---

[57] Hogan (2020:133).

[58] Lest the reader might find this preposterous we hasten to add that what we think Kant saw was the limitation of conceptual analysis for accounting for mathematical proofs. But we, contra Friedman and others, decouple conceptual analysis and logic in Kant.

[59] See Beth (1956–57) and (1957). How closely Hintikka follows Beth is problematic. This issue is discussed in Peijnenburg (1994).

synthetic should be accounted for in terms of inferential steps.[60] Thus, whether a proposition counts as analytic or synthetic will be derivative on the nature of the inferential steps that are needed to prove it.[61] The distinction between analytic and synthetic inferential steps is drawn within the boundaries of contemporary first order logic. Steps which introduce new individual constants (such as those needed in deploying existential elimination) are synthetic. Those which do not introduce any new constants are analytic. Notice that this distinction could be drawn even within a fragment of first order logic such as monadic first order logic.

Starting from the distinction Kant draws between concepts and intuitions, with the former characterized as general representations and the latter as individual representations, Hintikka[62] claims that what is essential to Kant's notion of intuition, as for instance in passage A717 cited above, is that it functions almost as a singular term for denoting an individual.[63] The connection between synthetic steps and Kant's construal of mathematical knowledge as synthetic a priori (thus grounded on intuition) rests on Hintikka's thesis that the syntheticity of mathematics in Kant's sense is based on the use of inferential rules that introduce new objects.

On this interpretation, a method or procedure is analytic if in it we do not introduce any new geometrical entities, in brief, if we do not carry out any constructions. A procedure is synthetic if such constructions are made use of, i.e. if new geometrical entities are introduced into the argument.[64]

Synthetic steps are those in which new individuals are introduced into the argument; analytic ones are those in which we merely discuss the individuals which we have already introduced.[65]

---

[60] In doing so, Hintikka gave pride of place to the methodological ("constructional") reading of the analytic/synthetic distinction as opposed to that based on concept containment or on epistemic features. The methodological reading finds its origin in Greek mathematics and in Pappus. See Hintikka and Remes (1974).

[61] See van Benthem (1974) for an excellent discussion of some major problems with Hintikka's attempt to connect analyticity, resp. syntheticity, for propositions with concepts of analytic, resp. synthetic, proofs.

[62] This is the result of a number of exegetical and theoretical moves on Hintikka's part; for instance, the historical claim that the Transcendental Aesthetic is posterior to the doctrine of method and that the connection of intuition to sensitivity is not essential in characterizing intuition, whose main feature is that of being an individual representation. Both claims have been widely contested. For Hintikka's contributions on this issue, see Hintikka (1967, 1969, 1973, 1974, 1978, 1981, 1984, 1992, 2020).

[63] "For Kant, an intuition is simply anything which represents or stands for an individual object as distinguished from general concepts" (Hintikka 1974: 130). Capozzi (2020) argues against the thesis that in Kant singular terms stand for intuitions.

[64] Hintikka (1973: 203).   [65] Hintikka (1973: 210).

We have here an example of an interpretation of Kant's thought that stands directly opposed to that of Russell. According to Hintikka's interpretation, the essence of Kant's thought does not consist in some sort of observation of the diagram, but rather rests in a specific form of inference involving individuals (akin to our existential elimination) that Kant considered to be mathematical. This is of course a very strong thesis, for it implies that even in ordinary reasoning in natural language when we proceed by quantifier elimination (going from, say, 'everything is identical to itself' to 'Fido is identical to itself') we are carrying out inferences that are characteristic of mathematics.[66]

It is in light of this rational reconstruction of Kant that Hintikka provides his explanation of B14. By connecting Kant's notion of demonstration to the *apodeixis* described by Proclus as the verification that the construction really accomplished what it was set out to do, Hintikka says:

> I suspect that a particularly perplexing passage in the first *Critique* receives a natural explanation pretty much in the same way as the remarks on arithmetic. I mean the statement Kant makes in B 14 to the effect that all inferences (*Schlüsse*) of the mathematicians are based on the principle of contradiction "which the nature of all apodeictic certainty requires." This passage becomes very natural if we take Kant for his word and understand him as referring solely to the apodeictic or 'proof proper' part of a Euclidean proposition. Taken literally, the proof proper or *apodeixis* is after all the only part of a Euclidean proposition where inferences are drawn. Taken in this way Kant's statement expresses precisely what he would be expected to hold on my interpretation, viz. that the distinction between on one hand *apodeixis* and on the other hand *ecthesis* and the auxiliary construction separates the analytic and the synthetic parts of a mathematical argument.[67]

---

[66] Hintikka is explicit about this: "What we are accustomed to call the logic of quantification in its general form would not have been logic at all for Kant, for according to Kant logic dealt with general concepts only. Quantificational theory, Kant would have been forced to say, hinges on non-logical, 'intuitive' methods. Of course, saying this is not much more than another way of saying that typically quantificational modes of inference would have been called by Kant mathematical rather than logical" (Hintikka 1973: 140). And he comments on a similar example to the one mentioned in the text as follows: "It is important to realize, for instance, that the inference from 'everyone is mortal' to 'Socrates is mortal' might very well be conceived of as being synthetic in the sense Kant had in mind. For him, a logical inference dealt with general concepts only. The introduction of an intuition made an argument synthetic. Now the term 'Socrates' is presumably a representative of an individual, and hence an intuition in Kant's sense. Therefore the inference, however trivial it might be, might very well be classifiable as a synthetic inference in Kant's sense of the term" (Hintikka 1973: 194). Capozzi (2020: 123) challenges Hintikka's claim that the inference from 'everyone is mortal' to 'Socrates is mortal' is synthetic; she does so by questioning that for Kant singular terms stand for intuitions.

[67] Hintikka (1974: 173).

Hintikka claims that "the use of *ecthesis* was for Kant a typically mathematical method of reasoning" and that "he could not use it in logic."[68] According to this reading, Kant's analysis of a Euclidean proposition would consist of a synthetic part, which we consider logical nowadays but that for Kant would have been typically mathematical but not logical, followed by an analytic part, which would be logical both in our sense and in Kant's sense. One conse-quence of this is that for Kant logic would be much weaker than monadic first order logic, a part of logic that admits of existential elimination and universal generalization.[69]

Hintikka's interpretation has failed to persuade many scholars. For instance, de Jong (1998) questions the connection Hintikka wants to establish between the constructional meaning of analysis/synthesis (as opposed to the directional meaning of it) and the Kantian definition of analytic and synthetic proposi-tions. The majority of commentators (Carson, Parsons) have objected to Hintikka's cavalier attitude towards the "immediacy" of intuition and van Benthem had pointed out some very serious issues concerning the various definitions of analyticity proposed by Hintikka.[70]

But one might disagree with Hintikka's general interpretation while still accepting that it highlights something crucial for our project. The majority of mathematicians and philosophers who provided a syllogized version of Euclidean theorems only focused on the *apodeixis* of the Euclidean proposi-tion. By contrast, Kant is also worried about giving an account of the con-struction steps appearing in a Euclidean proposition (or problem). Thus, for

---

[68] Hintikka (1974: 175). Hintikka tries to ground his thesis on what Kant says about reduction of syllogisms to those of the first and second figure in the 1762 essay *The False Subtlety of the Four Syllogistic Figures* (see Kant 1992b: 84–105). However, the inaccuracies of Hintikka's reading of that essay have been pointed out in Capozzi (1973: 252–6). We provide the following as a recent statement by Hintikka of the opposition in Kant between quantificational reasonings in mathematics and syllogistic reasonings in logic: "On the one hand, Kant's notion of logic was extraordinarily narrow, comprised essentially only of syllogistic reasoning. On the other hand, he was perfectly familiar with a plethora of first-order logical inferences, including plenty of applications of reasoning by instantiation in mathematics. However, they were not listed at his time under the title of 'logic.' They were the inferences one finds in traditional axiomatically developed elementary geometry, especially the way in which geometrical propositions were presented by Euclid and by many other traditional mathemati-cians" (Hintikka 2020: 86).

[69] Hintikka seems to oscillate on this issue. Speaking of Kant's position in the midst of a discussion on surface vs. depth tautology he says: "[Kant's position] consisted in restricting the field of logical truths to monadic quantification theory (traditional syllogistic), and in classifying all the other quantificational modes of reasoning as mathematical" (Hintikka 1973: 189–90). Since monadic pred-icate logic includes *ekthesis* and other forms of quantificational inference, this is in contrast to Hintikka's other claims about *ekthesis* and other quantificational forms of reasoning as lying outside Kant's logic.

[70] See Carson (2009), Parsons (1992), van Benthem (1974).

us, the preliminary question to focus on is: would Kant have agreed that the *apodeixis* of a Euclidean proposition only requires general logic (and thus, arguably, only syllogistic logic)? Hintikka suspects this to be the case. He says:

> Of course, if the introduction of new geometrical objects by constructions is disregarded, then the rest of a geometrical argument is completely analytic or, if you prefer, tautological. But this point is entirely traditional. It does not contradict Kant or Aristotle in the least. For Kant, the proof proper [*apodeixis*] was obviously analytic. I suspect he thought it could in principle be accomplished by means of syllogisms, just as some of his predecessors [Hintikka mentions Leibniz and Wolff in note] had maintained that all geometrical proofs can be accomplished in this way.[71]

We are now back to our questions Q1 and Q2. From the last quote we see that Hintikka thinks that for Kant the *apodeixis* of a Euclidean demonstration can be carried out by syllogisms. Moreover, according to Hintikka, this squares well with Kant's recognition that the entire proof, including *ekthesis*, cannot be carried out without more substantial quantificational resources and thus, Q1 is answered positively and Q2 negatively by Hintikka.

Our take with respect to Q1 and Q2 is as follows. We are, just like Hintikka, convinced that Kant would agree that the *apodeixis* can be carried out by syllogistic reasoning alone. Thus, we agree on the answer to Q1. Of course, here syllogistic theory is construed generously as to include modus ponens, disjunctive and conditional syllogism. But it will also have to include non-quantificational reasonings on basic relations (such as equality, congruence and even collinearity of points) that are not monadic relations. Would Kant also include in logic the forms of argument, such as *ekthesis*, used to reduce syllogisms to the first figure? Hintikka thinks not but this has been disputed by Kantian scholars.[72] But let us see if we can make some progress without getting entangled in this issue.

Hintikka's interpretation rests on the neat division in a Euclidean proof between a synthetic part and an analytic part. Should it turn out that the analytic part requires after all quantificational reasoning then Hintikka's interpretation will lose much of its appeal. To test whether this might be a threat to the interpretation, let us look again at the proof of Euclid I.1 as reconstructed by Mueller and then we will look at the *apodeixis* in more detail:

---

[71] Hintikka (1973: 218).   [72] See Capozzi (1973).

Protasis On a given finite straight line to construct an equilateral triangle.

Ekthesis Let *AB* [see Figure 1.2] be the given finite straight line.

Diorismos Thus it is required to construct an equilateral triangle on the straight line *AB*.

Kataskeuē With center *A* and distance *AB* let the circle *BCD* have been described; again with center *B* and distance *BA* let the circle *ACE* have been described; and from the point *C* in which the circles cut one another to the points *A*, *B* let the straight lines *CA*, *CB* have been joined.

Apodeixis Now since the point *A* is the center of the circle *CDB*, *AC* is equal to *AB*. Again, since the point *B* is the center of the circle *CAE*, *BC* is equal to *BA*. But *CA* was also proved equal to *AB*; therefore each of the straight lines *CA*, *CB* is equal to *AB*. And things which are equal to the same thing are equal to one another; therefore *CA* is also equal to *CB*. Therefore the three straight lines *CA*, *AB*, *BC* are equal to one another.

Sumperasma Therefore the triangle *ABC* is equilateral; and it has been constructed on the given finite straight line. Which was required to be done (Q.E.F.).

We have already seen that several scholars before Kant turned the *apodeixis* into a set of syllogisms. But when one looks at their reconstruction one immediately sees that the *apodeixis* contains a mix of general and particular sentences that require quantificational inferences. Consider the claim "since the point *A* is the center of the circle *CDB*, *AC* is equal to *AB*." How is this obtained? The construction has simply given the circles *CAE*, *CDB*, points *A*, *B*, *C*, and the segments *AC* and *AB*. To argue that *AC* is equal to *AB* we need the universal statement that in any circle any two radii are equal. From there (using already an instantiation to speak of the center *A* of an arbitrary circle) one uses universal instantiation to obtain *AC* = *AB*. Another universal instantiation is needed to prove *BC* = *BA*. The further inference leading to *CA* = *CB* requires the use of common notion 1 to the effect that if two quantities are equal to the same quantity then they are equal among themselves. The last step (therefore the three straight lines *CA*, *AB*, *BC* are equal to one another) does not require any universal instantiation.

As we have seen,[73] scholastic and late scholastic logicians expanded the traditional theory of syllogism to include a principle analogous to that of existential generalization, the principle of exposition for universal affirmative

[73] Cf. Chapter 2.

sentences and even identities. If we accept these integrations and call 'syllo-gistic' the resulting doctrine, it could seem plausible that the *apodeixis* in the proof just examined *has* a syllogistic form. It suffices a closer look to the *apodeixis*, however, to see that this is not the case, for the form of the premises cannot be easily fit to match the required form: the 'common notion,' for instance, as Thomas Reid in the eighteenth century would have said "cannot be brought into any syllogism in figure and mode."[74]

We have seen that Hintikka is willing to say that universal instantiations of this kind are synthetic for Kant. But then one cannot claim that the *apodeixis* is analytic or a fortiori (if all syllogistic inferences are analytic inferences) syllogistic. It thus become imperative to contextualize, and Hintikka is aware of this, the quantificational rules. Whether a *new* object is introduced by a rule such as universal instantiation will depend on whether that object has already appeared before in the proof. In our case the applications of universal instan-tiation in the *apodeixis* satisfy the condition that the instantiated objects have already appeared in the construction stage and thus are not new. Thus, Euclid I.1 successfully passes the test of conforming to Hintikka's reconstruction. The natural question to raise is whether this model can work for all of Book I of Euclid's *Elements*. We will come back to it after we discuss Friedman.

Here is the upshot of the analysis so far. According to Hintikka's analysis, the *apodeixis* of the Euclidean proofs (at least in book I of the *Elements*) is analytic according to his own reconstruction. Since he is offering his notion of analytic as a "rational reconstruction" of the historical Kant, he also must hold that the *apodeixis* is logical according to the historical Kant. We are less sanguine about the adequacy of Hintikka's notion of the analytic to capture the notion of logical inference in Kant but we nonetheless concord with the individual claims that (a) the *apodeixis* is analytic according to Hintikka's notion; and (b) Kant believed that the *apodeixis* was logical according to his notion of logical (which includes both propositional and syllogistic inferences; we have no doubt that he was also convinced that the presence of relational predicates could have been recast in form of monadic predicates).

While often associated with that of Beck, Brittan's position (see Brittan 1978) is closer to the Russellian line. The passages usually adduced to classify Brittan alongside Beck are the following. They both occur as comments to B14.

> Kant contends, to the contrary, that although in a mathematical proof the
> steps follow one another analytically, that is, in accordance with the principle

---

[74] Reid (1852: 701–2).

of contradiction, the initial premises of the proof and the conclusion are themselves synthetic.[75]

To refer back to a passage already cited, Kant comments on his predecessors as follows at B14: "For as it was found that all mathematical inferences proceed in accordance with the principle of contradiction . . . it was supposed that the fundamental propositions of the science can themselves be known to be true through that principle. This is an erroneous view." It could not be stated more clearly that all mathematical inferences are analytic. The synthetic character of the propositions of mathematics is a function of some feature of the propositions themselves and not of the way in which they come to be established.[76]

It is, however, misleading to use these passages as expressing Brittan's considered position. At that stage of the argument, Brittan is using B14 to claim that there are substantial textual hurdles confronting Hintikka's interpretation. Brittan's position has some points in common with that of Hintikka. For instance, he holds that in a mathematical proof there are elements of syntheticity (requiring appeal to intuition) that are provided by what we now call logic but that Kant would not have recognized as logic. His interpretation is, however, more radical than Hintikka's, and in some ways opposed to it. Brittan defends the idea that Kant's theoretical commitments are better captured by postulating that his logic is a universally free logic (he seems to espouse what he calls the Lambert–Parsons reconstruction).

We are now ready to describe one of the most influential interpretations of Kant's philosophy of mathematics in the last thirty years. Michael Friedman has put forward a reading of Kant's philosophy of mathematics which is centered on an interpretation of the role of pure intuition in Kant's system. Friedman starting point consists in deploying contemporary logical insights in order to clarify what is going on in Kant's work. In particular, Friedman emphasizes the connection of Kant's philosophy of mathematics to the Aristotelian logic which was predominant at the time. Friedman begins by remarking on one of the most salient features of Kant's theory:

What is most striking to me about Kant's theory, as it was to Russell, is the claim that geometrical reasoning cannot proceed analytically according to

[75] Brittan (1978: 46).   [76] Brittan (1978: 55).

concepts – that is, purely logically – but requires a further activity called "construction in pure intuition."[77]

Commenting on Kant's passages in A715–17/B743–5, Friedman claims that whereas we think of a proposition like Euclid 1.32 as following directly from the axioms by pure logic, for Kant intuition has to accompany the process of demonstration making a mathematical proof into a spatio-temporal object (this is much stronger than saying that the axioms require an a priori pure intuition in order to be grounded):[78]

Euclid's axioms do not imply Euclid's theorems by logic alone. Moreover, once we remember that Euclid's axioms are not the axioms used in modern formulations and, most important, that Kant's conception of logic is not our modern conception, it is easy to see that the claim in question is perfectly correct. For our logic, unlike Kant's, is polyadic rather than monadic

---

[77] Friedman (1992: 56).

[78] A classical claim concerning the essential role of the diagram and the constructive steps in Euclidean proof, which is perfectly analogous to Friedman's claim, is found in Schleiermacher. Schleiermacher in his *Dialektik* is engaged in showing that "the syllogism does not deserve the place that people have wanted to assign to it. It is not a third form to be added to concept and judgment but always only analysis" (Schleiermacher 1839: 289). In the context of arguing for this claim, Schleiermacher discusses the role of syllogism in mathematics and argues that despite appearances, in a mathematical proof the epistemic burden, given by the "Beweis," rests squarely on the constructions of auxiliary lines, whereas the syllogism can only be of use in the analysis of the construction itself:

When in recent times prominent reference is made to the mathematical procedure, — a procedure by means of which knowledge is obviously obtained — then it admittedly seems as if everything here is being done by means of syllogistic form. But this is not so; rather, everything comes down to the discovery of auxiliary lines. As soon as I have the latter, I already have the proof and am only in need of an analysis; whoever comes up with the construction merely analyses it afterward by means of the syllogism. The theorem can only emerge according to a heuristic or in a structural [architektonisch] way, not by accident: therefore, it cannot be that the figure exists without the proof existing with it. Good mathematicians do not care for the syllogism, rather they bring everything back to intuition. And so it now becomes clear that the syllogism does not deserve the place usually attributed to it. (Schleiermacher 1839: 288–9)

This dense passage reverberates with Kantian terminology. To the mathematical Beweis [proof], Schleiermacher opposes the final analysis (what classically would have been called the *Apodeixis*). The proof [Beweis], characterized as a construction, uses auxiliary lines, and it is the source of knowledge, which rests on intuition, and is therefore (even though Schleiermacher only leaves this implicit) the source of the syntheticity of the proof. The diagram and the proof [Beweis] go hand in hand.

A reply to Schleiermacher—whose lectures on *Dialektik* were rather influential among German logicians in the mid-nineteenth century—that emphasizes the importance of syllogisms in the extension of knowledge even in mathematics is found in Ueberweg's *System der Logik und Geschichte der logischen Lehren*. Ueberweg (1857) gave a proof of the parallel axiom based on 15 syllogisms of which 13 in Barbara. The proof is based on the notion of a direction of a line and it is part of a large variety of such proofs given in the eighteenth and the nineteenth centuries (see Mancosu 2016: chapter 2). Ueberweg's position is analyzed in Peckhaus (1999) where it is, however, erroneously claimed that the last editions of Ueberweg's text do not have the proof in question. We have checked all the editions (up to the fifth edition included) and the proof is found in all of them.

(syllogistic); and our axioms for Euclidean geometry are strikingly different from Euclid's in containing an explicit, and essentially polyadic, theory of order.[79]

Friedman goes on to characterize what are essential differences between the two logics. The essential one is that in monadic logic every (consistent) sentence can be made true by finitely many objects whereas in polyadic logic it is easy to express the existence of an infinite number of objects.[80] According to Friedman, monadic logic only enters in the final part of the Euclidean proof, after Euclid generates the necessary points by a process of construction. There are three major constructions:

(i)   draw a line segment between any two points
(ii) extend any line segment by any given line segment
(iii) draw a circle with any given point as center and any given line segment as radius.

The set of points that can be so constructed is $Q(\sqrt{p})$, that is, the closure of Q (the rationals) under square root.

In the contemporary axiomatization à la Hilbert, essential use of polyadic logic is involved in the axioms for '<' (dense linear order without endpoints). More essentially, in some of these axioms (for instance that yielding the property of density of the order <) we have a quantifier dependence "for all there exists." According to Friedman, Kant cannot represent the idea of infinity formally or conceptually. If logic is monadic one can only represent such infinity intuitively by an iterative process of spatial construction. Since the procedure of generating points, lines, and circles is a way to eliminate existential axioms Friedman concludes that for Kant "this procedure of generating new points by the iterative application of constructive functions takes the place, as it were, of our use of intricate rules of quantification theory such as existential instantiation" (65). Since the methods involved go well beyond

---

[79] Friedman (1992: 58–9).

[80] There are a number of inaccurate statements in Friedman's characterization of monadic logic. They have to do with Friedman's implicit assumption that one is working with finitely axiomatizable theories with finitely many basic predicates. For instance, on p. 68 he says that in monadic logic "the number of primitive predicates is always finite," which is definitely not the case. Or on p. 59 he says "Proof-theoretically, therefore, if we carry out deductions from a given theory using only monadic logic, we will be able to prove the existence of at most 2^k objects." Again, this is only true if we restrict the axiomatic theory in various ways, none of which depend essentially on the nature of monadic logic. In the remaining discussion we will not insist on these issues and interpret Friedman in a charitable way.

the essentially monadic logic available to Kant, Friedman views the inferences as synthetic rather than analytic.

What is, then, the picture that emerges from Friedman's account with respect to Kant's position on the issue that is relevant to us? We think it can be summarized as follows. First, according to this picture, Kant is concerned with the entire Euclidean proof, not just the *apodeixis*. One does not understand the synthetic nature of mathematical propositions if one does not see the elements of syntheticity that undergirds the proof in establishing a geometrical theorem. The syntheticity affects not only the judgment itself (i.e., the theorem as a synthetic a priori judgment) and the axioms from which the proof takes its start (which are also synthetic a priori) but more importantly it is the very essence of the "functional" constructions that drives the proof to its final conclusion. The synthetic steps are connected to the generation of individuals obtained by composition of the functional constructions corresponding to the construction postulates in Euclid. However, in a passage on pp. 86–7, Friedman seems to grant that the *apodeixis* of a Euclidean proof can be carried out syllogistically (or in monadic logic). In this he seems to agree with Hintikka's sentiment about the logic needed in the *apodeixis*. Let us look at this passage, warning the reader that we are correcting at one point a typo in the text.

> Now this conception of the role of calculation and substitution in mathematical proof also applies, *mutatis mutandis*, to the case of geometry. In Euclidean geometry we start with an initial set of basic constructive functions: the operation $f_L(x, y)$ taking two points x, y to the line segment between them, the operation $f_E(x, y)$ taking line segments x, y to the extended line segment of length $x + y$, and the operation $f_C(x, y)$ taking point x and line segment y to the circle with center x and radius equal to y. We also have a specifically geometrical equality relation (congruence) and, of course, definitions of the basic geometrical figures (circle, triangle, and so on). Euclidean proof then proceeds somewhat as follows. Given a figure a satisfying a condition...a..., we construct, by iteration of the basic operations, a new constructive function g yielding an expanded figure $g(a)$ satisfying a condition ---$g(a)$---. From this last proposition we are then able to derive a new condition __a__ on our original figure a. Whereas the inference from ---$g(a)$--- to __a__ [reading __a__ in place of...a...in the original text] can be viewed as "essentially monadic," and is therefore analytic or logical for Kant, the inference from...a...to ---$g(a)$---is not: it proceeds synthetically, by expanding the figure a as far as need be into the space around it, as it were.   (Friedman 1992: 86–7)

Thus, even on the Hintikka and the Friedman interpretation, it is claimed that the *apodeixis* for Kant is a mere matter of logical deduction.[81] As we pointed out in analyzing Hintikka, there are some residual worries about whether this is in fact the case when we interpret logic as being monadic logic. But we won't make heavy weather of that since it is clear from what Friedman says in various places that being "essentially monadic" for him just means the absence of quantifier dependence in a piece of reasoning.[82] We can now conclude this section by summarizing again the main points of contention concerning the issue we are after.

We began by distinguishing two questions. The first was whether Kant thought, like Wolff, that the entire Euclidean proof could be syllogized. We saw that the answer to this question is highly contested. On the Russell–Hintikka–Friedman variety of interpretations the answer is no. Kant introduces the intuition a priori as a synthetic element in the proof to "compensate" for the lack of logical resources. On the Couturat–Beck view, no appeal to a priori intuition is needed in the proof; the a priori intuition plays a role only in establishing the axioms. With the axioms on board, the proof can be carried out purely logically (even though the logic required, as we now know but unbeknownst to Kant, would have to go beyond the confines of Kant's logic). By contrast, concerning our second question, whether the *apodeixis* can be carried out purely logically (even in Kant's sense), both interpretations agree that it can.

We would like now to mention the importance of some recent work on the formalization of Euclidean geometry for some of the above topics. If one looks for contemporary formalized systems of Euclidean geometry that might capture some of the intuitions we have been discussing, we can first of all ask whether the general picture painted by Hintikka and Friedman is a plausible reconstruction of Euclid's practice in Book I of the *Elements* (we restrict discussion to Book I as some further issues would have to be discussed if we wanted to include the remaining books of the *Elements*; and Book I is in any case all that gets discussed in connection to Kant). One complication in approaching the topics by means of contemporary formal systems is that the formal system could either be a quantifier-free system or a system with a quantificational apparatus (such is that set up in Beth 1957, where Beth gives a detailed reconstruction of I.32 in connection to discussing Kant's theory of

---

[81] It is worth remarking that Hintikka locates the non-logical aspects of the proof in the *ekthesis* while Friedman locates it in the *kataskeuē*.

[82] See note 13 of Friedman (1985: 466) and notes 8, 14, and 21 of Friedman (1992: 63, 65, 71).

mathematics). In the first case, we are thinking of a formalization given in terms of Skolem functions capturing the required Euclidean constructions. One can simplify things by taking "point" as the only primitive non-logical predicate (as in Tarski's axiomatization of geometry). In such a set up one can only express universal quantifications by means of free-variables; all existential statements are instanced by terms obtained using the Skolem functions. In the second case, we allow a logic with quantifiers but then we must recover the partition between a purely constructive part of the Euclidean proof (corresponding to the *ekthesis* and the *kataskeuē*) and the logical part (corresponding to the *apodeixis*) by means of a metatheorem showing that given a statement S in Euclid's Book I and a proof P of the statement in the formalized system, one can find a proof P* of S such that the first part consists only of propositional moves and the construction of terms and the second part consists only of quantifier steps. One way of doing so is through a sequent calculus axiomatization such as that offered by Beeson and using the mid-sequent theorem for the system (see below).

The remarkable thing about the formal reconstructions of Euclid that have been offered in the last decade is that several things have become clear on this front. First of all, the complexity of all the statements in Euclid, Book I, is at most of the form $\forall x \exists y A(x, y)$ where $A(x, y)$ is quantifier-free. This greatly simplifies the logical situation in that, on account of the simple logical structure of the axioms that can be chosen to axiomatize the system, various theorems from proof theory yield exactly the possibility of transforming derivations in the sense indicated above. We will mention first of all the system of Skolemized continuous Tarski geometry studied by Beeson.[83] Here the set up is that of a quantifier-free theory where every construction is expressed by a Skolem function. Universal generalizations are expressed by provable sentences with free variables. This formalization corresponds quite well to the Friedman interpretation of Kant: every Euclidean proof can be structured so that all the needed points are constructed first and then the theorem itself is given in terms of $A(x_1, \ldots x_n, f(x_1, \ldots x_n))$. The term $f(x_1, \ldots x_n)$ corresponds to the existential that we would normally write expressing the theorem as $\forall x_1, \ldots x_n \exists y A(x_1, \ldots x_n, y)$. Beeson (2015) also offers a different formalization without Skolem functions but simply based on first-order logic; the system is formulated in a sequent calculus style.[84] Unlike

---

[83] Beeson (2015); see also Beeson (2022).
[84] In Beeson's systems the axioms have been modified so that the points they assert to exist are unique and depend continuously on parameters.

the previous formalization in terms of Skolem functions we can have sentences with quantifiers (say $\forall x \exists y\ A(x, y)$) but then we need a metatheorem showing that any proof P of a statement S can be transformed into a proof P* of S, in which the introduction of all the terms precedes the introduction of the quantifier steps.[85] This is obtained by applying the mid-sequent theorem, (or Herbrand's theorem, itself a consequence of cut-elimination) to the system. The system of classical continuous Tarski geometry, presented in Beeson (2015), has exactly this metatheoretical property.[86]

What this shows is that the general reconstruction provided by Hintikka and Friedman is actually a coherent reconstruction of the Euclidean practice and that if Kant had seen things as they describe he would have been espousing a reading of the *Elements* that contemporary axiomatic investigations vindicate. Of course, even though Beeson tries to stay as close to Euclid as possible much had to be changed to get a fully rigorous system. In particular the role of equality, betweenness, and collinearity as basic relations is crucial in showing that the quantifier free parts of the Euclidean reasoning in the *apodeixis* cannot be monadic. But we have already pointed out that Friedman has a rather extended notion of monadic logic, which would be better captured by a logic lacking quantifier alternations (and thus allowing binary or n-ary relations as being part of "monadic" logic).

Euclid I.1 asks to construct an equilateral triangle on a given finite straight line.[87]

The formalized statement has the usual $\forall x \exists y A(x,y)$ form, namely:

$$\forall A \forall B\ (A \neq B \rightarrow \exists X(ABX \text{ is a triangle} \& ABX \text{ is equilateral})).$$

Of course, being a triangle and equilaterality are defined in terms of the more basic relations of betweenness and collinearity. The formalized proof is more complicated than the original Euclidean proof also on account of the fact that Euclid does not verify that ABX is a triangle (i.e., that A, B, and X are not collinear).

Similar results can be obtained looking at the system E set up by Avigad, Dean, and Mumma (2009). System E was conceived as a hybrid system to represent the interaction of text and diagram in Euclid. In E there are rules (or

---

[85] See Beeson (2015: 1263).
[86] Beeson is very interested in intuitionistic versions of these theories but this aspect is irrelevant for our concerns.
[87] The reader who is curious to see how a fully formalized proof of Euclid I.1 looks like in Beeson's system of classical continuous Tarski's geometry can consult Beeson, Narboux, and Wiedijk (2019: 247–8).

derived rules) that can be naturally understood as Skolem functions. These are the rules that are applied in what corresponds to a Euclidean construction in an E derivation. It is important to emphasize that there is no general background logic in E. What's derived are lists of literals in a form that corresponds to a statement of the form $\forall x \exists y (A(x) \rightarrow B(x,y))$ in a first order theory. This form is primitive in E. It is not defined via quantifiers and connectives that can be iterated definitely to form new sentences. E does not have quantifiers and connectives. The reasoning that then follows in the *apodeixis* is according to E very simple in a logical sense. It consists (mostly) in the application of rules which in a first-order theory of geometry would appear as universal statements of the form

$$\forall x(A(x) \rightarrow B(x))$$

where $A(x)$ and $B(x)$ are conjunctions of literals with free variables x. When a proof does not consist entirely of the application of such rules in E, it involves a case analysis. Suppose $C(x)$ is the list of literals assumed or derived about a configuration in an E derivation and $L(y)$ is an atomic E relation whose variables y are among the variables x of $C(x)$. One can then split the derivation into two cases, where one case is given by the list of literals

$$C(x), L(y);$$

and the other is given by the list of literals

$$C(x), \text{not } L(y).$$

If one can derive the same list of literals $C'(x)$ from all case branches generated in this way, then one has proven $C'(x)$ according to E. This rule, coupled with E's *ex-falso* rule, is how E models Euclid's reductio proofs.

Even in this case, Friedman's interpretation requires an extended notion of monadic reasoning. Indeed, the literals of E are atomic relations or negations of atomic relations (e.g., X is between Y and Z, X and Y are on the same side of line Z, XY = ZW). There are no one-place predicates in E. But, as we have now established, not everything has to fit into Aristotelian syllogisms with three monadic terms for it to be monadic according to Friedman.

Where does this leave us with respect to Kant? Our sense is that Kant would definitely have espoused the thesis that the *apodeixis* of a Euclidean proof is syllogistic. The presence of binary relations in the apodeixis (equality of segments) would not have stopped him from doing that as, in line with a long list of previous logicians, he would have been persuaded that only a little massaging would have been needed to recast all such sentences into categorical sentences using monadic (in the *strict* sense of monadic) predicates.

We have seen that the major lines of interpretation agree that Kant would have considered the *apodeixis* of a Euclidean proof as purely logical. They disagree, however, on whether the construction part of the Euclidean proof (*ekthesis* and *kataskeuē*) would have been considered purely logical by Kant. What gives us pause in accepting the Friedmanian line is a major assumption in Friedman's interpretation, namely that there is an exact match in Kant between conceptual analysis[88] and monadic logic (even in the extended sense he entertains of this notion). If one decouples analysis and monadic logic, then it is possible to claim that what Kant had seen was not the limitation of his logic for capturing mathematics but rather the limitation of (his rather impoverished notion of) analysis for capturing mathematics and that this is what led him to postulate an a priori intuition.

This solution could still be consistent with a version of the Couturat–Beck view in that Kant could still be able to think that his logic can recapture the entire Euclidean proof. He would be wrong in that belief (as we now know) but in making that assumption he would not be setting himself apart from what the majority of logicians up to him (and many after) simply thought, namely that by appropriately massaging the form of the premises and conclusions, any kind of recalcitrant statement could be expressed in subject-predicate form. A consequence of this would of course be that of claiming that for Kant some logical inferences would not be analytic according to his notion of analysis.[89] The point could also be made without appeal to inferences: some monadic validities would not be classified as analytic truths. For instance, with A and B monadic, "every A is A or B" although monadic is certainly not analytic according to the criterion of concept inclusion. Note that biting the bullet and saying that A or B is contained in A is very unpalatable since one can show, iterating disjunctions with new disjuncts, that the intension of a concept can have infinitely many distinct concepts, something which Friedman himself says Kant denies.[90] A recent paper questioning whether Kant thinks of the truths of logic as analytic is Heis (forthcoming). But we will stop here as the

---

[88] By conceptual analysis we mean here the process of decomposition of the concepts into their marks.

[89] We say this fully aware that perhaps one should restrict, *apud* Kant, talk of analyticity to judgments without extending it to inferences. There are, however, some passages in Kant where the extension is contemplated but our point does not require the extension of the analytic/synthetic distinction to inferences and can be, as we do in the main text, stated in terms of judgments rather than inferences.

[90] See his discussion of the intension and extension of a concept in Kant on p. 67 of Friedman (1992).

dialectic becomes very subtle and pursuing things further would take us beyond any reasonable amount of space for this section.

One final comment before we close this section on Kant. The importance of Kant in the overall story we are recounting cannot be exaggerated. According to Russell (and other commentators), Kant was led to introduce the notion of a priori intuition because he realized that the logic of his time could not account for mathematical demonstrations in Euclid. This reading has been extremely influential and in that sense Kant played, and continues to play, a very important role in our story. However, our reading of what led Kant to introduce an priori intuition rests on the claim that what Kant realized was the limitation of his notion of conceptual analysis for accounting for Euclidean proofs. According to our reading, he still believed (wrongly as it turns out), in line with many of his contemporaries, that the logic of his time could, with appropriate transformations of the sentences, account for all valid inferential patterns, including those used in mathematics. If one grants this decoupling of conceptual analysis and formal logic, then Kant's historical position appears less innovative with respect to the story we are recounting. But given the enormously influential Russellian reading that does not make Kant's role in the overall story we are recounting any less important.

7

Bernard Bolzano on Non-Syllogistic Reasoning

Bernard Bolzano devoted quite a bit of attention to the issue of non-syllogistic reasoning even though he does not seem to have spent much time worrying about oblique inferences. However, his discussion of non-syllogistic inferences was, from the beginning of his career, framed within a general discourse on mathematical method and it is thus intimately related to our topic.

In *A better grounded presentation of mathematics* (1810),[1] the topic is addressed after Bolzano claimed that he thought all syllogisms were reducible to *Barbara*. However, he went on to assert that he believed there were other types of non-syllogistic inferences:

> I believe that there are some *simple kinds of inference* [*Schlußarten*] apart from the syllogism. I want to point out briefly those which have occurred to me so far.
>
> (a) If one has the two propositions:
>
> *A is* (or *contains*) *B*, and
>
> *A is* (or *contains*) *C*
>
> then by a special kind of inference there follows from these the third [proposition]:
>
> *A is* (or *contains*) [*B et C*].
>
> This proposition is obviously different from the first two, each considered in itself, for it contains a different predicate. It is also not the same as their *sum*, for the latter is not a *single* proposition but a collection of *two*.[2]

The context in which Bolzano introduced such inferential forms was related to his theory of ground and consequence. While all the inferential forms he is

---

[1] For a recent evaluation of the importance of this booklet, see Rusnock and Sebestik (2013).
[2] Russ (2004, 110–11).

Syllogistic Logic and Mathematical Proof. Paolo Mancosu and Massimo Mugnai, Oxford University Press.
© Paolo Mancosu and Massimo Mugnai 2023. DOI: 10.1093/oso/9780198876922.003.0008

discussing are valid inferential forms not all of them, as it will appear below, are examples conforming to ground and consequence:

> Finally, it is also obvious that according to the necessary law [*Gesetz*] of our thinking the first two propositions can be considered as *ground* for the third and not, indeed, conversely.
>
> (b) In the same way it can also be shown that from the two propositions,
>
> *A is* (or *contains*) *M*, and
>
> *B is* (or *contains*) *M*
>
> the third proposition:
>
> [*A et B*] *is* (or *contains*) *M*
>
> follows by a simple inference.
>
> (c) Again, it is another simple inference, which, from the two propositions
>
> *A is* (or *contains*) *M*, and
>
> (*A cum B*) *is possible* or *A can contain B*
>
> derives the third:
>
> (*A cum B*) *is* (or *contains*) *M*.[3]
>
> This inference has a great deal of similarity to the syllogism but is nevertheless to be distinguished from it. The syllogistic form would really arrange the second of the two premisses thus: *(A cum B) is* (or *contains*) *A*. But this proposition first needs, for its verification [*Bewährung*], the proposition, *(A cum B) is possible*. But if this is assumed, one can dispense with the first as being merely analytic, or take it in *such* a sense that, in fact, both mean the same thing and only differ verbally. All these kinds of inference, including the syllogism, have the common property that from two premisses they derive only one conclusion [*Folgerung*]. On the other hand, the following might look like an example of a kind of inference whereby two conclusions come from *one* premiss:

---

[3] Bolzano does not explain here what he means by "et" and "cum." Rusnock and Sebestik provide a useful gloss (Rusnock and Sebestik 2013: 8):

> Bolzano then introduces new rules of inference involving the conjunctions et and cum. He does not explain the meaning of the concepts designated by 'et' and 'cum' in the first issue of the Contributions, but does so in the unpublished second installment (Bolzano 1977: §32ff.). According to what he says there, 'et' represents ideal combination, which is possible between any two concepts, while 'cum', by contrast, represents real combination, which is not always possible. The concepts "circle" and "square", for example, can be combined ideally (since one can think of a circle along with a square), but not really (since, as he then maintained, one cannot even form the concept of a circle which is square). One expedient that seems to work fairly well is to read 'et' as 'along with' or 'as well as' and 'A cum B' as 'A, which is B'.

A *is* (or *contains*) [B *cum* C]
therefore
A *is* (or *contains*) B, and
A *is* (or *contains*) C.

But I do not believe that this is an *inference* in that sense of the word which we established at the beginning of this section. Having recognized the truth of the *first* of these three propositions I can, indeed, *recognize* subjectively the truth of the two others, but I cannot view the first *objectively* as the *ground* of the two others. I cannot let myself go into a detailed discussion of all these assertions here.[4]

While Bolzano's interest seems to have been focused on the ground-consequence relation, we should not overlook the importance of his claims concerning the existence of valid inferences that are not syllogistic. Moreover, the existence of such valid patterns of reasoning is not peripheral and is present in Bolzano's mind when he is tackling various metatheoretical claims, as for instance his claim that "for every simple concept there is at least one unprovable judgment in which it appears as subject."[5] In that discussion he explicitly takes into consideration the existence of non-syllogistic valid arguments and he reiterated:

Now I have already mentioned in §12 that I do not regard the syllogism as the only simple kind of inference.[6]

Since these matters are discussed in a treatise on a better-grounded presentation of mathematics, there is no problem in inferring that for Bolzano there are, also in mathematics, valid forms of reasoning—some of which preserve the ground-consequence relation—that are not syllogistic. What is surprising, however, is that Bolzano does not bother with discussing any of the canonical examples (Euclid I.1, I.32, or I.46) that were usually adduced as evidence that all of mathematics could be recast in syllogistic fashion. Bolzano explicitly remarks on the existence of non-syllogistic inferences in note 1 to section 155 of the *Wissenschaftslehre* (1837) devoted to the notion of derivation [*Ableitung*]. But it is in section 262 of the *Wissenschaftslehre*, titled "The syllogism in the received logic," that Bolzano provides a rigorous argument aimed at showing that proofs (including mathematical proofs) cannot all be

---

[4] Russ (2004: 110–11).   [5] Russ (2004: 117–18).   [6] Russ (2004: 118).

captured by chains of syllogistic inference rules. This argument is obviously very important for our topic and we accordingly analyze it in what follows.

The section begins by stating the novelty of Bolzano's opposition to the traditional claim that, apart from immediate inferences, all inferences in logic are of syllogistic nature:

> It is assumed in almost all previous treatises of logic that besides the immediate inferences, concerning the existence of which not all logicians agree, there is only one single kind of inference which, through combination and repetition, gives rise to all the others. This single kind of simple inference is supposed to consist essentially of three propositions, in which three ideas that are considered variable are divided in such a way that one of them occurs in each of the two propositions which are regarded as premises of the third, while the other variable ideas in these premises both occur in the conclusion. From the most ancient times, this kind of inference has been called a syllogism; in recent times, they have also been called inferences of reason. It is worth the trouble to examine the ways people tried to convince themselves that there was no other kind of simple yet mediate inference than the one I just described.[7]

Given Bolzano's extensive knowledge of the previous logical literature, it is surprising that he does not give credit to Jungius and other logicians who preceded him in pointing out the insufficiency of syllogistic reasonings. It is true that some of them construed oblique reasonings as immediate inferences but Bolzano's silence is nonetheless surprising.

Bolzano begins his discussion with the broad notion of syllogism given by Aristotle in *Prior Analytics* I.1:

> A syllogism is a discourse in which, certain things being stated, something other than what is stated follows of necessity from their being so.[8]

Bolzano expresses astonishment that Aristotle, having started with such a broad definition, restricted its range quite narrowly. A possible justification for the restriction consists in two claims, made by Aristotle at I.25 and I.23 of *Prior Analytics*, to the effect that every inference can be reduced to a sequence

---

[7] Bolzano (2014: 430).

[8] In Aristotle (1989), Smith uses "deduction" for "syllogism." While we will cite from the Smith edition we have preferred to keep "syllogism" in this first quote to avoid confusion.

of syllogistic arguments, each of which consisting of two premises and one conclusion. In *Prior Analytics* I.25, Aristotle tries to argue the thesis by relying on a lemma that he claims to have already established:

> For suppose that E has been concluded from A, B, C and D. Then some one of them must have been taken in relation to another, one as whole and the other as part (for this was proved earlier, viz., that if there is a deduction it is necessary for some of the terms to be related in this way).[9]

Bolzano objects that it is questionable whether Aristotle ever gave such a proof. For this reason, he focuses on the claim made in section 23, for if the claim made in section 23 can be correctly established then the thesis asserted in section 25, according to Bolzano, will follow. Before we embark on Bolzano's analysis of section 23, we should mention that several commentators (Corcoran, Lear, Smiley) see the claims made by Aristotle in sections I.23 and I.25 as attempts to establish the reducibility of all inferences to syllogistic inferences.[10] For instance, Jonathan Lear claims that "In *Prior Analytics* I.23 and I.25 he [Aristotle] argues that every deductive argument can be expressed as a series of syllogistic inferences."[11] And later, comparing the ambiguity of the word 'syllogism,' which admits of a broad and a narrow interpretation, to that of 'deduction' in contemporary use, he says:

> This ambiguity is tolerable since the value of the formal syllogistic is supposed to derive from the fact that a syllogism in the broad sense can be represented as a syllogism in the narrow sense. [...] Aristotle is declaring that any non-formal deduction, such as the proof that a triangle has interior angles equal to two right angles, can be recast as a formal deduction. One might call this claim 'Aristotle's Thesis.'[12]

It is thus against Aristotle's thesis that Bolzano argues in section 262 of the *Wissenschaftslehre*:

---

[9] Aristotle (1989: 39; I.25, 42a5–10).

[10] We will not raise the issue, quite present in Aristotle and in the contemporary secondary literature, as to how to account in this context for hypothetical syllogisms and for syllogisms by impossibility.

[11] Lear (1986: ix).

[12] Lear (1986: 11). We remark here that Lear (1986: 12–13) refers to a long passage from Pr. An. I.35 and claims that in that passage Aristotle offers an example of a mathematical theorem (the sum of the internal angles of a triangle is two right angles) that cannot be reduced to a syllogistic argument. We will not discuss this further but we only indicate that, even according to Lear, this passage cannot be reconciled with Pr. An. I.23.

I would rather examine the way in which he [Aristotle] proves in chapter 22 [*sic* for 23] that every inference occurs in one of the three (Aristotelian) figures.[13]

The Aristotelian passage from I.23 is the following:

Now, if someone should have to deduce A of B, either as belonging or as not belonging, then it is necessary for him to take something about something. If, then, A should be taken about B, then the initial thing will have been taken. But if A should be taken about C, and C about nothing nor anything else about it, nor some other thing about A, then there will be no deduction (for nothing results of necessity through a single thing having been taken about one other). Consequently, another premise must be taken in addition. If, then, A is taken about something else, or something else about it or about C, then nothing prevents there being a deduction, but it will not be in relation to B through the premises taken. Nor when C is taken to belong to something else, that to another thing, and this to something else, but it is not connected to B: there will not be a deduction in relation to B in this way either. For, in general, we said that there cannot ever be any deduction of one thing about another without some middle term having been taken which is related in some way to each according to the kinds of predications. For a deduction, without qualification, is from premises; a deduction in relation to this term is from premises in relation to this term; and a deduction of this term in relation to that is through premises of this term in relation to that. And it is impossible to take a premise in relation to B without either predicating or rejecting anything of it, or again to get a deduction of A in relation to B without taking any common term, but <only> predicating or rejecting certain things separately of each of them. As a result, something must be taken as a middle term for both, which will connect the predications, since the deduction will be of this term in relation to that. If, then, it is necessary to take some common term in relation to both, and if this is possible in three ways (for it is possible to do so by predicating A of C and C of B, or by predicating C of both A and B, or by predicating both A and B of C), and these ways are the figures stated, then it is evident that every deduction must come about through some one of these figures. (The argument will also be the same if A is connected with B through more things: for the figure will be the same even in the case of many terms.)[14]

---

[13] Bolzano (2014: vol. II, 431).   [14] Aristotle (1989: 36–7; tr. Smith).

In order to understand Bolzano's objections, we need to spell out Aristotle's argument. A reconstruction of the argument must, however, keep in mind that Aristotle subscribes to what Smiley has called the "chain principle," "namely the principle that the premises of a syllogism must form a chain of predications linking the terms of the conclusion" (see, for instance, An. Pr. 40b30ff. and 41a6). We will follow here the enlightening reconstruction of Aristotle's argument given in Paoli (1990).[15]

We first define a formal language AL and specify at the same time the metavariables.

Variables: A_1, A_2, A_3, ...

Metavariables $(A, B, C, ...)$

Terms of AL: if A is a variable then A is a term; nothing else is a term.

Metavariables for terms: $t, s, ...$

Formulas: if t and s are terms, then ts is a formula. Nothing else is a formula. Thus, AB is a formula and its intuitive meaning is the Aristotelian "B belongs to A"

Metavariables for formulas; b, a, a_1, a_2, ...

Furthermore, Γ and Δ will be used as metavariables ranging over finite multisets of formulas. Recall that in a multiset the same element might appear more than once.

A demonstration is a pair $<\Gamma, b>$ where b is different from each element of Γ. Informally the set Γ represents the premises and b the conclusion of the demonstration

Σ, Σ' etc. are metavariables for demonstrations.

With the formal language in place, here is the reconstruction of the Aristotelian theorem given by Paoli.

Theorem. Every proof Σ of $b = AB$ has one of the following forms: $<\{CB, AC\}, AB>$, $<\{BC, AC\}, AB>$, $<\{CB, CA\}, AB>$.

Proof:

Let Σ be a demonstration of AB from Γ, i.e. $<\Gamma, AB>$. Γ is non-empty and AB is not in Γ by assumption. Thus, there is a formula a in Γ that expresses that a term belongs to another. Since the conclusion is AB there must be a common variable between a and AB. We have the following three possibilities:

(1) $A = AB$: in that case Σ is not a demonstration because the conclusion cannot appear as one of the premises;

---

(2) $A = AC$ (resp. CA): We have two cases to consider.

Case I: if $\Gamma = \{AC\}$ or neither A nor C appear in Γ-$\{AC\}$ then, by the chain principle, no conclusion follows.

Case II: Either A or C or both, but not B, occur in Γ-$\{AC\}$ one can have a demonstration in which Γ is the set of premises but B, by the chain principle, will not occur in the conclusion.

(3) $A = BC$ (resp. CB): this is analogous to the case $a = AC$ (resp. CA)

In order to infer AB two premises are thus necessary and from the above it is clear that A and B must each occur in some premise but not in the same. Let $a_i \neq a_j$ be the two premises in Γ. In order to infer BA there must be a connection between A and B in a_i and a_j. By the chain principle the only way for this to happen is for a_i and a_j to have in common a variable C different from A and B. But it is easy to see that there are only three combination of premises that satisfy the requirement: $\{CB, AC\}$, $\{BC, AC\}$, $\{CB, CA\}$.

Bolzano raises a number of objections to Aristotle's argument. The first objection concerns the form of the conclusion "B does or does not have A":

> This form results if it is presupposed that every proposition that is to be proved should have neither more nor less than two variable ideas (A and B), where one of these is the complete subject idea, the other the complete predicate idea. But this is not at all the case; often we have to prove propositions in which only a single idea is to be viewed as variable and frequently we have to prove others where three or more ideas are simultaneously considered variable. These variable ideas are distributed in many different ways: sometimes they are in the subject, sometimes in the predicate idea, sometimes they form these by themselves, and at other times in conjunction with other ideas.[16]

According to Bolzano, the restriction to two variable ideas (one for the subject and one for the predicate) is unjustified. By way of examples, Bolzano mentions hypothetical syllogisms (with one variable idea) and disjunctive syllogisms (with arbitrary finite number of variable ideas). Incidentally, in §264 Bolzano categorically dismisses disjunctive syllogisms from qualifying as syllogisms.

---

[16] Bolzano (2014: vol. II, 432).

The second objection concerns Aristotle's assumption, displayed in the previous reconstruction, that "a proposition which leads to the conclusion that *A* is or is not *B* must necessarily contain *A*." Bolzano gives as counterexample an argument in which "neither of the premises has either a subject or a predicate idea which is the same as the subject or predicate idea of the conclusion":

Caius plays the flute
Titus plays the organ
Caius and Titus are two distinct persons
Thus, among Titus and Caius there is a flute player and an organ player.

Bolzano's third objection concerns the claim, also used in the above reconstruction, that from one premise nothing follows. But Bolzano remarks that all immediate inferences yield a conclusion based on a single premise.

The fourth objection rejects Aristotle's claim that in the conclusion of a derivation there cannot be a term that does not already appear in one of the premises. He considers GA and DA. By Aristotle's claim it would follow that B has A cannot be deduced. But this, objects Bolzano, does not follow. By letting B be $G+D$ one concludes that B has A.

Finally, and this is the fifth objection, there are inferences in which one can introduce a variable idea that was not present in the premise. Bolzano's example is:

A has objectuality (i.e., A is not empty).
Thus, at least one of 'Some A are B' and 'Some A are not B' is true.

But Bolzano is not done yet. He considers two more attempts which aim at showing that every inference is syllogistic, the first by Krug (1806) and the second by Maaß (1793). He pinpoints where their arguments beg the question. In his discussion, Bolzano asserts that from the general notion of argument follows neither the restriction to two premises nor that to a single conclusion. But even though all the attempts he criticizes (including Aristotle's) are unsuccessful one cannot conclude that the thesis is false:

> Now the proofs might be deficient, but the assertion itself, that the syllogistic form of inference is the only one, could nevertheless be true. Hence it is necessary to show more explicitly than has already been done that there are many forms of inference which differ essentially from syllogisms.[17]

---

[17] Bolzano (2014: vol. II, 435).

The sense in which this is the case is spelled out in stages. First, Bolzano starts with a broad definition of argument in which multiple conclusions are allowed.

> If we assume that every inference, whether syllogistic or not, can be put in the form of the following proposition, "Every collection of ideas whose substitution for $i, j, \ldots$ makes propositions $A, B, C, D, \ldots$ true also makes propositions $M, N, O, \ldots$ true", and if we are furthermore willing to call every proposition which has this form an inference, then there is no doubt that there are inferences with any desired number of premises as well as conclusions. For, any number of propositions $A, B, C, D, \ldots$, so long as they are compatible with respect to ideas $i, j, \ldots$ can function as premises, and any number of propositions $M, N, O, \ldots$, so long as they become true whenever the former are true (and a great number of these can always be found), can be envisaged as conclusions. Among the propositions $M, N, O, \ldots$ there may also be some which do not require the truth of all of the propositions $A, B, C, D, \ldots$, but only of one or several of them.[18]

He provides an example of such an inference with multiple conclusions (*All As are Bs, All Cs are Ds; thus, Some As are Bs, and some Cs are Ds*) but then he objects that those who claimed that every inference consists of three propositions would have objected to his broad definition, and to the specific example. They would have objected on account of the fact that

(a) they had in mind an inference with a single conclusion and
(b) the argument should have no more premises than are required for the derivation of the conclusion (in other words, there might be redundant premises).

Thus, the focus becomes more restricted, and Bolzano discusses inferences that he calls "exact" (see section 155 (entry 26) of the *Wissenschaftslehre*), that is, which do not fall prey to objections (a) and (b). Is it possible that all exact inferences are syllogistic in nature? The first example discussed is a *Sorites*, which has the form:

A is *B*
B is *C*

---

[18] Bolzano (2014: vol. II, 435).

C is D

.

.

L is M

A is M

However, since a *Sorites* can be derived by a reiterated application of ordinary syllogisms, Bolzano's case for non-syllogistic reasoning would be rather weak if that is all he had. But that is not the case and his claim is much stronger:

> However, I should like to claim that this is not the case; *there are forms of inference which tell us how to derive certain propositions from others which cannot be derived from them by any syllogisms, no matter how often they are repeated.* To be convinced of this important truth, nothing is required but a review of the various inferences which were discussed in this part, and to try to derive the indicated conclusions from their premises through one or several syllogisms. I wish to record only a few examples.[19]

Let us look at the examples.

a. Consider the following two premises:

Every A is B

Every B is A.

And now consider the conclusion: Thus, every object which stands under one of the ideas A or B also stands under both of them.

Bolzano comments: "This conclusion can never be derived by ordinary syllogistic methods; rather, the latter lead us to the identical proposition 'Every A is A.'"

b. Consider n propositions $A_1, A_2, \ldots, A_n$ for which it holds:

There is only one A_1

There is only one A_2

...

There is only one A_n

No object that falls under A_i falls under A_j for $i \neq j$ ($1 \leq i, j, \leq n$)

Thus, The number of $A_1, A_2, \ldots, A_n$ is n.

---

[19] Bolzano (2014: vol. II, 436).

Therefore, Bolzano asks: "through what single syllogism or combination of several of them could this conclusion be derived?"

c. Consider the argument (with exclusive 'or'):
Either *A* or *B*
Either *A* or *C*
Either *A* or *B* or *C*
Thus, *A* is true.

In the final part of §262, Bolzano discusses a possible maneuver to block his argument which consists in claiming that all such inferences are incomplete syllogisms. The idea is that each of these inferences must be replaced by a syllogism with a hypothetical major premise condensing the rule to the effect that the conclusion follows from the premises. To illustrate the maneuver on a simple example, consider *modus ponens*:

P
If P then Q
Q.

The reformulated version would say:

$(P\&(P \rightarrow Q)) \rightarrow Q$
$P\&(P \rightarrow Q)$
Thus, Q.

We had already encountered a discussion of such strategy in Hoffmann. Bolzano rejects the move with the following justification:

> We must note, however, that the major premise in this inference is itself a complete inference; if this inference has any usefulness, and if its validity can be judged merely from logical concepts, then it must itself be discussed in logic. Hence nothing was gained by this expedient.[20]

We consider the above sufficient evidence for claiming that Bolzano defended the idea that mathematical proofs cannot be entirely recaptured syllogistically. His claim in the 1810 essay is made in the context of a logic construed as

---

[20] Bolzano (2014: vol. II, 437).

mathematical method and in the *Wissenschaftslehre* the claim for the existence of non-syllogistic proofs extends to the realm of all the sciences. There is in addition a section in the *Wissenschaftslehre*, §530, where Bolzano more explicitly comments on mathematical proofs. The context is that of converting every indirect proof into a direct one and Bolzano does it using as an example a proof of Euclid (I.19) and claiming that he has successfully done it for all proofs by contradiction in Euclid. He says:

> It has always been recognized that apagogical proofs in my sense do not indicate the objective ground of the proposition proved. E. Reinhold (L., p. 409) would not even give them the name of logical proofs. Aristotle (Anal. prior., II, 14) attempts to demonstrate that every apagogical proof can be converted into an ostensive one. Leibniz, on the contrary (Nouv. Ess., IV, viii) is of the opinion that this is at least difficult in many cases. Here I would like to side with Aristotle. For even though I think his proof is inadequate (*being based upon the presupposition that all inferences are in only one of the three syllogistic figures* [our emphasis]), I have succeeded in every case I tried (namely, with all the proofs per absurdum in Euclid's Elements) in converting apagogical into ostensive proofs.[21]

What is important here is the diagnosis for why Aristotle's argument is not successful: he tried to prove the result only using syllogistic reasoning. But those inferences are not sufficient to recapture the transformation of an indirect proof into a direct proof.

Incidentally, this section of the *Wissenschaftslehre* remains, in our opinion, the best evidence for claiming that Frege had read Bolzano.[22] Frege in *Logic in Mathematics* reduces every indirect proof to a direct one using exactly the same strategy, and the same example from Euclid (I.19), discussed by Bolzano.

Let us conclude this section by raising an obvious question. Did Bolzano's work as a mathematician play a role in his realization that there are non-syllogistic inferences? We wish we had a satisfactory answer to this question. Unfortunately, Bolzano's arguments do not seem to appeal specifically to insights he would have gained through his experience as a working mathematician. Rather, they seem to be motivated by the analysis of the logical claims defended by Aristotle and others. We cannot go beyond this conclusion but it is certainly an issue worth investigating further.

---

[21] Bolzano (2014: vol. IV, 208).   [22] See Mancosu (1996: 117).

At this point, we will leave Austria and Germany and move to the United Kingdom and head toward the epilogue of our narrative. We will then examine, respectively, an 'opponent' (Thomas Reid) and a defender (William Hamilton) of the syllogism and, after a brief 'interlude' involving Mill, we will deal, finally, with Augustus De Morgan, who will be shown to have a twofold attitude toward the syllogism: on the one hand, he considers it inadequate for proving theorems in mathematics, and on the other hand, while introducing substantial modifications that radically change its structure, he continues to regard it as a fundamental logical tool.

8

Thomas Reid, William Hamilton, and Augustus De Morgan

This chapter is mainly devoted to discussing Augustus De Morgan's ideas on syllogism and its utility in proving mathematical (geometrical) theorems. Before analyzing De Morgan's attempt to prove the so-called Pythagorean theorem, however, we will present some views of Thomas Reid, William Hamilton, and John Stuart Mill concerning the traditional (Aristotelian) syllogism: this short detour is motivated by the requirement of putting De Morgan's inquiries in their historical context.

Thomas Reid (1710–1796)—a philosopher known as the founder of the so-called Scottish school of *common sense*—had a poor opinion of traditional (Aristotelian) logic. He devoted an entire essay of about one hundred pages to explain and discuss Aristotle's logic and his concluding remark was that although "the art of categorical syllogism is better fitted for scholastic litigation, than for real improvement in knowledge, it is a venerable piece of antiquity, and a great effort of human genius" (Reid 1806: 135). Therefore, Aristotelian logic deserves our attention in the same way as do the ancient relics of human history, the pyramids of Egypt or the Chinese wall:

We admire the pyramids of Egypt, and the wall of China, though useless burdens upon the earth. We can bear the most minute description of them, and travel hundreds of leagues to see them. If any person should, with sacrilegious hands, destroy or deface them, his memory would be had in abhorrence. The predicaments and predicables, the rules of syllogism, and the topics, have a like title to our veneration as antiquities: they are uncommon efforts, not of human power, but of human genius; and they make a remarkable period in the progress of human reason.[1]

---

[1] Reid (1806: 135; 1822: 123).

Syllogistic Logic and Mathematical Proof. Paolo Mancosu and Massimo Mugnai, Oxford University Press.

According to Reid, logic by itself, insofar as it is built on the basis of the Aristotelian theory of syllogism, does not guarantee any progress "of useful knowledge":

> The slow progress of useful knowledge, during the many ages in which the syllogistic art was most highly cultivated as the only guide to science, and its quick progress since that art was disused, suggest a presumption against it; and this presumption is strengthened by the puerility of the examples which have always been brought to illustrate its rules. The ancients seem to have had too high notions, both of the force of the reasoning power in man, and of the art of syllogism as its guide.[2]

Reid thinks that "mere reasoning," without observation and "experiments properly conducted" can be only of very little help in most subjects: "[T]he power of reasoning alone, applied with rigour through a long life, would only carry a man round, like a horse in a mill, who labours hard, but makes no progress."[3] For Reid, mathematics, not logic, is the only field where the exercise of "mere reasoning" is truly rewarding:

> The relations of quantity are so various, and so susceptible of exact mensu-ration, that long trains of accurate reasoning on that subject may be formed, and conclusions drawn very remote from the first principles. It is in this science, and those which depend upon it, that the power of reasoning triumphs: in other matters its trophies are inconsiderable.[4]

To learn how to reason, one must study mathematics.

The categorical syllogism, according to Reid, has its foundation in a prin-ciple stating that "what is affirmed or denied of the whole genus may be affirmed or denied of every species and individual belonging to it." This is a principle of "undoubted certainty"—so states Reid—"but of no great depth."[5] Aristotle and all the logicians after him, assumed it as an axiom, as the starting point of the entire syllogistic system, which, "after a tedious voyage, and great expense of demonstration" lands at last in it, as its ultimate conclusion.[6] Thus, Reid sharply distinguishes logic (the Aristotelian syllogistic in particular),

---

[2] Reid (1806: 87–8; 1822: 107).   [3] Reid (1806: 88; 1822: 107).
[4] Reid (1806: 88; 1822: 107).   [5] Reid (1806: 94; 1822: 109).
[6] Reid (1806: 94; 1822: 109). For an analogous statement, see (1806: 82; 1822: 105):

> As to the legitimate modes, Aristotle, and those who follow him the most closely, demon-strate the four modes of the first figure directly from an axiom called the *Dictum de omni et*

from mathematics: only this latter, not the former, is productive of knowledge. Concerning the importance of mathematics in strengthening the powers of reasoning, Reid fully agrees with Locke:

> I agree with Mr. Locke, that there is no study better fitted to exercise and strengthen the reasoning powers, than that of the mathematical science; for two reasons: first, because there is no other branch of science which gives such scope to long and accurate trains of reasoning; and secondly, because in mathematics there is no room for authority, or for prejudice of any kind, which may give a false bias to the judgment.[7]

With great clarity, Reid connects the problem of relations and, implicitly, of the so-called *oblique terms* with the structure of propositions, and firmly states that it is false that every proposition has 'subject-predicate' form. Because relations play a fundamental role in mathematics, he concludes that the four categorical sentences of traditional syllogistic logic are unfit to develop proofs containing relations, and therefore that syllogism is a useless engine in mathematics:

> If it should be thought that the syllogistic art may be a useful engine in mathematics, in which pure reasoning has ample scope: First, it may be observed, That facts are unfavourable to this opinion: For it does not appear that Euclid, or Apollonius, or Archimedes, or Huygens, or Newton, ever made the least use of this art; and I am even of opinion that no use can be made of it in mathematics. I would not wish to advance this rashly, since Aristotle has said, that mathematicians reason for the most part in the first figure. What led him to think so was, that the first figure only yields conclusions that are universal and affirmative, and the conclusions of mathematics are commonly of that kind. But it is to be observed, that the propositions of mathematics are not categorical propositions, consisting of one subject and one predicate. They express some relation which one quantity bears to another, and on that account must have three terms. The

---

nullo. The amount of the axiom is, that what is affirmed of a whole genus, may be affirmed of all the species and individuals belonging to the genus; and that what is denied of the whole genus, may be denied of its species and individuals. The four modes of the first figure are evidently included in this axiom. And as to the legitimate modes of the other figures, they are proved by reducing them to some mode of the first. Nor is there any other principle assumed in these reductions but the axioms concerning the conversion of propositions, and in some cases the axioms concerning the opposition of propositions.

[7] Reid (1806: 130; 1822: 121).

quantities compared make two, and the relation between them is a third. Now, to such propositions we can neither apply the rules concerning the conversion of propositions, nor can they enter into a syllogism of any of the figures or modes. We observed before, that this conversion, A is greater than B, therefore B is less than A, does not fall within the rules of conversion given by Aristotle or the Logicians; and we now add, that this simple reasoning, A is equal to B, and B to C, therefore A is equal to C, cannot be brought into any syllogism in figure and mode.[8]

This passage is remarkable for at least 3 reasons:

(1) it suggests a reasonable answer to the question why Aristotle claimed that the most part of theorems in mathematics can be proved by means of first figure syllogisms;

(2) it recognizes that relations may contribute to the building of a proposition and that the elementary form of a proposition usually is not 'subject-copula-predicate';

(3) it states that it is impossible to transform some relational inferences— as, for instance, those asserting the transitivity of equality—into "syllogisms in figure and mode."

Sir William Hamilton, in his edition of Reid's posthumous works, commenting on Reid's criticism, retorted first, without any explanation, that the propositions mentioned by Reid were all categorical.[9] Then, concerning Reid's statement that the inference 'A is equal to B, and B to C, therefore A is equal to C' cannot take the form of a traditional syllogism, he remarked that this was due exclusively to the peculiar way in which the inference was expressed. According to Hamilton, indeed, it is a kind of enthymematic inference, in which a premise is suppressed:

Not as it stands;[10] for, as expressed, this reasoning is elliptical. Explicitly stated, it is as follows:

What are equal to the same, are equal to each other;

A and C are equal to the same (B);

Therefore, A and C are equal to each other.[11]

---

[8] Reid (1806: 90–1); see also Reid (1852: 701–2).   [9] Merrill (1990: 16).

[10] That is, the inference cannot be reduced in the form in which it is presented.

[11] Reid (1852: 702). As Ian Mueller remarks, an analogous solution was proposed by Alexander of Aphrodisia (third century AD) in response to Galen's criticism:

To reinforce his point and to show that it is possible to proceed syllogistically in demonstrating theorems in geometry and mathematics, Hamilton appeals to Herlinus and Dasypodius' work:

> Dr Reid could have found a rare work in the College Library of Glasgow, which it might have been profitable for him to consult—viz., an edition of the first six books of Euclid, by Herlinus and Dasypodius, in which every demonstration is developed in regular syllogisms. But this development did not render syllogistic what was not syllogistic from the beginning—it only shews that it was always so. A Reasoning is not the less syllogistic, because not formally enounced in two orderly premises and a conclusion. This, however, is the notion that many of those who have written about and against logic, seem to have entertained.[12]

As we have seen, however, it is surely an expression of unwarranted optimism to claim that in Herlinus and Dasypodius' book "every demonstration is developed in regular syllogisms."

It is worth remarking, however, that Hamilton's attitude toward syllogism was not too conservative: Hamilton, indeed, proposed to quantify the predicate of the categorical sentences traditionally composing a syllogistic figure. This means that he considered perfectly legitimate to include in a syllogism statements as, for instance, 'All men are some animals' or 'some violinists are some Europeans.' Hamilton was very proud of what he considered to be an authentic discovery that should improve the traditional syllogistic theory, and he became very upset when he discovered that even Augustus De Morgan had the same idea of quantifying the predicate. He accused De Morgan of plagiarism, thus giving birth to a bitter dispute that lasted for years and that was based on very shaky grounds, given that the idea of quantifying the predicate in a categorical statement was present in several authors before the nineteenth century. As we have seen, Vagetius, for instance, was one of these authors (and Leibniz expressed his approval of Vagetius on this point[13]).

Galen criticizes the Aristotelians for trying by force to count relational syllogisms as categorical. The subsequent discussion in the *Institutio*, supplemented with Alexander's logical commentaries, makes it virtually certain that the Peripatetic way of treating Galen's relational syllogisms was to add a universal premiss corresponding to Galen's axiom and to reformulate the argument as a 'categorical syllogism'. To give one example, Alexander transforms the unsystematically conclusive argument 'A is greater than B; B is greater than C; therefore A is greater than C' into:

> Everything greater than a greater is greater than what is less than the latter;
> A is greater than B which is greater than C;
> Therefore A is greater than C. (Mueller 1974: 62)

On 'Galen's axiom' and the Aristotelian syllogism, see Barnes (2007: 418–30).

[12] Reid (1852: 702).   [13] Cf., footnote 35 of chapter 4.

Hamilton's suggestion that the inference expressing the transitivity of equality needs to be completed with a general principle is in full agreement with Euclid's usual practice of invoking *common notions*, when proving theorems in the *Elements*.[14] Even so completed, however, the inference proposed by Hamilton is not a syllogism in the Aristotelian sense. To see why, let us consider again Reid's inference together with Hamilton's amendment:

Things that are equal to the same thing, are equal to each other;
A and C are equal to the same thing (B);
Therefore, A and C are equal to each other.

As David Merrill remarks in his book on De Morgan,

> It is not at all clear that the minor premise and the conclusion are categorical propositions at all, for each asserts that a certain pair – the pair CA and CB – has a certain property. It seems more natural just to say that CA is equal to CB; and it is not clear whether, say, traditional logic would countenance relations as properties of pairs.[15]

Moreover, we have to face the problem of how to justify, from the formal point of view, the second premise of the above argument. Thus, suppose it is inferred this way:

CA and AB are equal to one another;
CB and AB are equal to one another:
Therefore, CA and CB are equal to the same thing.[16]

This, however, is an inference with five terms; therefore, it cannot be a syllogism. As Merrill observes:

> Hamilton's formulation would only mask the problematic character of this inference by writing the original minor premise as, CA and CB are things equal to the same thing (AB), where there is no indication that their equality to the same thing must be inferred from their equality to AB. He might reply, of course, "That is obvious," but the point of logic is to capture the obvious.[17]

---

[14] Merrill (1990: 16).   [15] Merrill (1990: 17).   [16] Merrill (1990: 17).
[17] Merrill (1990: 17).

Other strategies can be pursued to logically derive the second premise but, as David Merrill has shown, the result does not change: there is no way of dressing in proper 'syllogistic clothing' Reid's inference. Because Merrill has investigated and explained with great clarity the reasons why Hamilton's reply to Reid does not work, we limit ourselves to refer the reader to the part of his book devoted to this issue and we now turn our attention to John Stuart Mill's *System of Logic*.

Even though John Stuart Mill in his *System of Logic* (1843) claims that any process of reasoning or inference is grounded on an inference from particulars to particulars "authorized by a previous inference from particulars to generals," he firmly defends the importance of syllogisms "for the purposes of reasoning":

> But while these conclusions appear to me undeniable, I must yet enter a protest, as strong as that of Archbishop Whately himself, against the doctrine that the syllogistic art is useless for the purposes of reasoning. The reasoning lies in the act of generalization, not in interpreting the record of that act; but the syllogistic form is an indispensable collateral security for the correctness of the generalization itself.[18]

Thus, he devotes an entire chapter of the *System* to describe the traditional theory of the syllogism, referring the reader to the more exhaustive presentation included in Richard Whately's *Elements of Logic*: "The reader may, however, be referred, for every needful explanation, to Archbishop Whately's *Elements of Logic*, where he will find stated with philosophical precision, and explained with remarkable perspicuity, the whole of the common doctrine of the syllogism."[19] And in the same vein as Whately, Mill states that all valid ratiocination, "all reasoning by which, from general propositions previously admitted, other propositions equally or less general are inferred" may be exhibited in syllogistic form. Consequently, Mill's opinion is that

> The whole of Euclid, for example, might be thrown without difficulty into a series of syllogisms, regular in mood and figure.[20]

According to the scholastic tradition, Mill too considers the *dictum de omni et nullo* as "the basis of the syllogistic theory."

---

[18] Mill (1843: 191; 1974: 196).   [19] Mill (1843: 168; 1974: 166).
[20] Mill (1843: 168; 1974: 166).

In chapter IV of his *Logic* ("Of trains of Reasoning and Deductive Sciences"), Mill demonstrates the fifth proposition of the first book of the *Elements* aiming to show how the inferences employed in the proof ultimately rest on inductive procedures. In this circumstance, however, he never attempts to convert the proof into a chain of syllogisms (or pseudo-syllogisms). As many other authors of the time, Mill, too, simply takes for granted that any mathematical proof is made by means of syllogisms. Moreover, he believes that "all correct ratiocination admits of being stated in syllogisms of the first figure alone."[21]

We now move on to chronicle De Morgan's attitude with respect to the reducibility of geometrical proofs to syllogisms. We will show that while De Morgan held fast to the claim that geometrical proofs were reducible to syllogisms, the claim itself underwent a subtle shift due to a mutation in De Morgan's account of syllogisms.

In line with the prevailing scholastic tradition, De Morgan considers the elementary structure of any sentence as composed of a subject, a predicate and the copula (the verb 'to be' in its various forms). Thus, as he specifies in his 1831 essay *On the Study and Difficulty of Mathematics* (henceforth *SDM*), a sentence like 'all right angles are equal' is composed of three different parts: the (quantified) subject ('(all) right angles'),[22] the predicate ('equal') and the copula connecting subject and predicate, "which is generally the verb *is*, or *is equal to*, and can always be reduced to one or the other."[23] In this essay, De Morgan claims that the traditional copula may be interpreted in two different ways: as corresponding to the predicative use of the verb 'is' on the one hand, or to 'is equal to,' on the other.[24] This is a remarkable novelty in a text that, in general, is quite conservative and conforms itself to the standard view according to which any theorem in Euclid's *Elements* can be demonstrated by means of syllogisms. To prove this last point, De Morgan, when speaking of *geometrical reasoning*, shows how the Pythagorean theorem can be demonstrated through a chain of syllogisms. We apologize for quoting here the entire passage, but we think that it will facilitate the comprehension of the remarks we will make later.

De Morgan introduces his demonstration of the Pythagorean theorem with the following words:

---

[21] Mill (1843: 244; 1974: 167).

[22] "the subject spoken of, viz., right angles, which is here spoken of universally, since every right angle is a part of the subject". De Morgan (1831; 1943: 203).

[23] De Morgan (1831; 1943: 203).   [24] On De Morgan's 'two copulas,' cf., Merrill (1990: 26–35).

The elements of geometry present a collection of such reasonings as we have just described, though in a more condensed form. It is true that, for the convenience of the learner, it is broken up into distinct propositions, as a journey is divided into stages; but nevertheless, from the very commencement, there is nothing which is not of the nature just described. We present the following as a specimen of a geometrical proposition reduced nearly to a syllogistic form. To avoid multiplying petty syllogisms, we have omitted some few which the student can easily supply.[25]

Interestingly enough, De Morgan acknowledges that his proof is not based on syllogisms in the proper sense of the word, but that it is reduced *nearly* to a syllogistic form and that some syllogisms have been omitted: he leaves to the students the task of supplying them.

The demonstration of the theorem runs as follows:

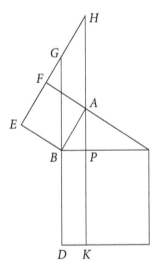

Hypothesis. *ABC* is a right-angled triangle the right angle being at *A*.

Consequence. The squares on *AB* and *AC* are together equal to the square on *BC*.

Construction: Upon *BC* and *BA* describe squares, produce *DB* to meet *EF*, produced, if necessary, in *G*, and through *A* draw *HAK* parallel to *BD*.[26]

Demonstration.

I. Conterminous sides of a square are at right angles to one another. (Definition.)
EB and BA are conterminous sides of a square. (Construction.)

---

[25] De Morgan (1831; 1943: 220).   [26] De Morgan (1831; 1943: 221).

∴ EB and BA are at right angles.

II. A similar syllogism to prove that DB and BC are at right angles, and another to prove that GB and BC are at right angles.

III. Two right lines drawn perpendicular to two other right lines make the same angle as those others (already proved); EB and BG and AB and BC are two right lines, etc., (I. II.).
∴ The angle EBG is equal to ABC.

IV. All sides of a square are equal. (Definition.)
AB and BE are sides of a square. (Construction.)
∴ AB and BE are equal.

V. All right angles are equal. (Already proved.)
BEG and BAC are right angles. (Hypothesis and construction.)
∴ BEG and BAC are equal angles.

VI. Two triangles having two angles of one equal to two angles of the other, and the interjacent sides equal, are equal in all respects. (Proved.)
BEG and BAC are two triangles having BEG and EBG respectively equal to BAC and ABC and the sides EB and
BA equal. (III. IV. V.)
∴ The triangles BEG, BAC are equal in all respects.

VII. BG is equal to BC. (VI.)
BC is equal to BD. (Proved as IV.)
∴ BG is equal to BD.

VIII. A four-sided figure whose opposite sides are parallel is a parallelogram. (Definition.)
BGHA and BPKD are four-sided figures, etc. (Construction.)
∴ BGHA and BPKD are parallelograms.

IX. Parallelograms upon the same base and between the same parallels are equal. (Proved.)
EBAF and BGHA, are parallelograms, etc. (Construction.)
∴ EBAF and BGHA are equal.

X. Parallelograms on equal bases and between the same parallels, are equal. (Proved.)
BGHA and BDKP are parallelograms, etc. (Construction.)
∴ BGHA and BDKP are equal.

XI. EBAF is equal to BGHA. (IX.)
BGHA is equal to BDKP. (X.)
∴ EBAF (that is the square on AB) is equal to BDKP.

XII. A similar argument from the commencement to prove that the square on AC is equal to the rectangle CPK.

XIII. The rectangles BK and CK are together equal to the square on BC. (Self-evident from the construction.)
The squares on BA and AC are together equal to the rectangles BK and CK. (Self-evident from XI. and XII.)
∴ The squares on BA and AC are together equal to the square on BC.

This demonstration is plagued with problems analogous to those that we have found in Herlinus and Dasypodius' proof.[27] Starting our analysis from the beginning of the proof, let us assume that the first premise of step

---

[27] De Morgan possessed a copy of Herlinus and Dasypodius' book; cf., De Morgan (1847: 286):

I am told that Dugald Stewart, who had a strong notion of the practical impossibility of presenting Euclid in a syllogistic form, never would believe that it has been done by

I (Definition) can be transformed into a standard categorical sentence. A problem arises, however, as soon as we move from the first to the second (minor) premise of the same step. The sentence "EB and BA are conterminous sides of a square" has two subjects (the sides EB and BA) and cannot be easily transformed into a sentence with only one subject without changing its meaning. We may try to split this sentence into *two* sentences, one with EB and the other with BA as subject, but then we need to transform the predicate 'being conterminous' (maybe employing synonymous expressions) in such a way that it can be applied to each subject in turn. But 'conterminous' is an expression that necessarily refers to two sides and this obstructs finding another expression that applies to only *one side* preserving the same meaning.

A similar remark applies to the conclusion "EB and BA are at right angles." This sentence too has two subjects and therefore does not have the form of a 'standard' categorical sentence. If we say something like 'EB is the side of a right angle' and 'BA is the side of a right angle,' we need to specify that EB and BA are both sides of the *same* right angle. In this case too, transforming the minor premise of the argument into a categorical sentence seems a hopeless task. Other cases of sentences with multiple subjects appear in the body of the demonstration and they all present the same difficulty.

Another difficult point is connected with the twofold meaning that in *SDM* De Morgan attributes to the copula. As David Merrill remarks, De Morgan in his demonstration treats equality "in two ways at once: first, as the predicate in 'All right angles are equal,' whose copula is 'is,' and then as part of the second copula 'is equal to.'"[28] Thus, in step IV, for instance, equality is treated as a predicate:

IV. All sides of a square are equal.
AB and BE are sides of a square.
∴ AB and BE are equal.

Herlinus and Dasypodius. Such a work is entered in catalogues: but I must say that the state of catalogues is such that Stewart or any one else had full right to doubt of any work, upon no other than catalogue evidence. The work does exist, and I have a copy of it. But seeing how matters stand, no one has right to declare that an old book ever was written, without informing his reader on what sort of evidence he relies.

The book is even mentioned in De Morgan (1868).
[28] Merrill (1990: 30).

But later on, in step VII, De Morgan treats 'is equal to' as a copula:

VII. BG is equal to BC.
BC is equal to BD.
∴ BG is equal to BD.

Still, Merrill observes that

> Our confusion about the two uses of equality is compounded by the fact that in moving from Step IX to Step XII, De Morgan infers "EBAF is equal to BGHA" from "EBAF and BGHA are equal," without comment.[29]

Another source of troubles is De Morgan's use of what Merrill characterizes as "multiple subsumption," as it is represented at step VI:[30]

VI. Two triangles having two angles of one equal to two angles of the other, and the interjacent sides equal, are equal in all respects.
BEG and BAC are two triangles having BEG and EBG respectively equal to BAC and ABC and the sides EB and BA equal.
∴ The triangles BEG, BAC are equal in all respects

The first premise states that if a triangle has two angles equal to two angles of another triangle and if the interjacent sides of the two triangles are equal, then the two triangles are equal 'in all respects.' The equality relation here holds first between two sides of two triangles, then between the interjacent sides of the same triangles and, finally, from all this the equality of the two triangles follows. The second (minor) premise presents an instance of two individual triangles that satisfy the conditions mentioned in the first premise (they are subsumed under the general case), so that the conclusion can be drawn. Giving to each of these premises the form of a categorical sentence is quite a hard task.[31]

A further critical remark made by Merrill concerns the minor premise of step IV. This premise is inferred by means of an argument which is missing in De Morgan's demonstration and which, as Merrill suggests, may be reconstructed as follows:

[29] Merrill (1990: 31).   [30] Merrill (1990: 31).
[31] Cf., Merrill (1990: 31): "While this has a semi-syllogistic form, it is in fact much more complex than that. In addition to the use of multiple subsumption, additional clauses must be added, such as that side EB is the interjacent side for BEG and EBG. Whether the fully articulated inference could be construed as merely a case of multiple subsumption is an open question."

AB and BE are sides of ABEF.

ABEF is a square.

∴ AB and BE are sides of a square.

This, however, "is not a syllogism in De Morgan's sense and, in fact, its validation requires the use of a principle which invokes embedding terms ('ABEF' and 'square') in a relational expression."[32]

Thus, if compared with other previous attempts to employ syllogisms to demonstrate geometrical theorems, De Morgan's demonstration of the Pythagorean theorem sounds quite traditional: De Morgan simply assumes without any further explanation that several different inferences can be reduced to syllogisms (as we have seen, the proof is partly left as an exercise to the reader).

Independently of the demonstration of the Pythagorean theorem, however, *SDL* contains some seeds of De Morgan's future interest for the logic of relations. Consider, for instance, how De Morgan, concluding the chapter on geometrical reasoning, characterizes the validity of an argument:

> In all that has gone before we may perceive that the validity of an argument depends upon two distinct considerations, 1) the truth of the relations assumed, or presented to have been proved before; 2) the manner in which these facts are combined so as to produce new relations, in which last the reasoning consists.[33]

This characterization is quite generic, but interestingly enough it does not appeal, as De Morgan will do in later publications, to the *form* of an argument to establish its validity: it refers, instead, to the truth determined by relations and to the 'production' of *relations*.

Eight years after the *SDL*, De Morgan published a small book entitled *First Notions of Logic*, which later became the first chapter of his *Formal Logic*, published the same year of Boole's *The Mathematical Analysis of Logic* (1847). *First Notions* is a quite peculiar book: it presents the syllogism in a traditional way, faithful to the scholastic and late-scholastic point of view, but it contains, at the same time, some interesting novelties. De Morgan begins *First Notions* claiming that 'A is B' and 'A is not B' are the elementary forms to which any proposition can be reduced:

---

[32] Merrill (1990: 31–2).   [33] De Morgan (1831; 1943: 228).

All propositions are either assertions or denials and are thus divided into affirmative and negative. Thus, A is B, and A is not B, are the two forms to which all propositions may be reduced. These are, for our present purpose, the most simple forms [...].[34]

De Morgan, however, is well aware that the reduction is not easy to be performed and admits that it "will frequently happen that much circumlocution is needed to reduce propositions" to the schematic form 'A is B.'[35] Oddly enough, as an example of proposition to be 'reduced,' he chooses the following:

"If he should come tomorrow, he will probably stay till Monday."

To justify the kind of 'reduction' he proposes, De Morgan remarks that in this proposition,

> There is evidently something spoken of, something said of it, and an affirmative connection between them. Something, if it happens, that is, the happening of something, makes the happening of another something probable; or is one of the things which render the happening of the second thing probable.[36]

Thus, the final result is:

A	is	B
The happening of his arrival tomorrow	is	An event from which it may be inferred as probable that he will stay till Monday.

Clearly, behind this 'reduction' there is not a general method susceptible of being applied to any kind of proposition which does not have the explicit form 'A is B,' and the final result, in this case at least, is quite cumbersome and naïve at the same time.

Confronted with the proposition 'A is less than B, B is less than C, therefore A is less than C,' De Morgan recognizes that "to write an ordinary syllogism" in a manner that shall correspond to it, some amount of "definite assertion" has to be introduced. Thus, he proposes three arguments that he considers to be necessary for representing the above proposition dressed in syllogistic (or quasi-syllogistic) 'clothes':

---

[34] De Morgan (1847: 4).   [35] De Morgan (1847: 4).   [36] De Morgan (1847: 4).

Every A is B, and there are Bs which are not As
Every B is C, and there are Cs which are not Bs
Therefore Every A is C, and there are Cs which are not As.

The Bs contain all the As, and more
The Cs contain all the Bs, and more
The Cs contain all the As and more.

From   Every A is B; [Some B is not A]
          Every B is C; [Some C is not B]
Follows   Every A is C; [Some C is not A].

De Morgan considers the third argument as expressed in 'the most technical form': however, it is not a syllogism. If we limit ourselves to consider the left half of it—that is, the inference composed of premises and conclusion not included in square brackets—then we have a proper syllogism. But if we want to construct an argument with a meaning even approximately similar to that of the above proposition, then each bracketed proposition has to be considered as joined to that preceding it. This means that each premise of the argument must be taken as a conjunction of two propositions.

Even though in *First Notions* De Morgan continues to believe that the syllogism is the pillar on which every mathematical demonstration rests, he recognizes that there are valid arguments not in syllogistic form (and which cannot be reduced to that form) that are usually employed to determine the premises of the syllogisms. As a prominent example of these arguments, De Morgan mentions the so-called *Dictum de omni*.

In his *Formal Logic*, De Morgan emphasizes the role played by the *dictum* for validating the traditional *syllogism ex obliquis*:

There is another process which is often necessary, in the formation of the premises of a syllogism, involving a transformation which is neither done by syllogism, nor immediately reducible to it. It is the substitution, in a compound phrase, of the name of the genus for that of the species, when the use of the name is particular. For example, 'man is animal, therefore the head of a man is the head of an animal' is inference, but not syllogism. And it is not mere substitution of identity, as would be 'the head of a man is the head of a *rational animal*' but a substitution of a larger term in a particular sense.

Perhaps some readers may think they can reduce the above to a syllogism. If *man* and *head* were connected in a manner which could be made subject and predicate, something of the sort might be done, but in appearance only. For example, 'Every man is an animal, therefore he who kills a man kills an animal.' It may be said that this is equivalent to a statement that in 'Every man is an animal; some one kills a man; therefore some one kills an animal,' the first premise *conditionally*, involve the conclusion as *conditionally*. This I admit: but the last is not a syllogism: and involves the very difficulty in question. 'Every man *is* an animal; some one *is* the killer of a man': here is no middle term. To bring the first premise into 'Every killer of a man is the killer of an animal' is just the thing wanted.[37]

Later on, in his second essay on the syllogism (1850), De Morgan associates to the substitutivity rule corresponding to the *dictum* a second rule consisting in substituting "a larger term used particularly for a smaller one":

A little consideration suggests as a necessary rule of inference, the right to substitute a larger term used particularly for a smaller one, however used, and a smaller, used in either way, for a larger used universally. What we may affirm or deny of *some* or *all* men, we may affirm or deny of some *animals*: what we may affirm or deny of *all* animals, we may affirm or deny of *all* or *some* men. The second part of the rule is the *dictum de omni et nullo*; the first part has not, within my reading, been added to it: both may well be incorporated in one under the name of the *dictum de majore et minore*.[38]

As we have seen when discussing Ockham's and Buridan's treatment of oblique inferences, the substitutivity rule corresponding to the *dictum* has the property of being *upward monotone*; the second rule mentioned by De Morgan, instead, has that of being *downward monotone*. Victor Sánchez Valencia has shown that these rules, even though not sufficient to validate oblique inferences, are on the right track: complementing them with the traditional theory of *distribution* (of the terms in a proposition) and with standard results concerning the composition of monotone functions, we may obtain a systematic procedure for validating oblique syllogisms.[39]

Thus, De Morgan admits that there are valid inferences that neither are nor can be reduced to syllogisms (those validated by the two substitution rules just mentioned, for instance). This conclusion forces him to admit that the traditional 'Aristotelian' syllogism does not occupy the center of all mathematical

---

[37] De Morgan (1847: 114).　　[38] De Morgan (1966: 28–9).
[39] Cf., Sánchez Valencia (2004: 494–503).

demonstrations. At the same time, however, he claims that the syllogism properly understood *is* the basic ingredient of any mathematical demonstration. 'Properly understood' means that the syllogism should be considered as an inference in which relations play a fundamental role. As we read in De Morgan's third essay on the syllogism (1858):

> When two objects, qualities, classes, or attributes, viewed together by the mind, are seen under some connexion, that connexion is called a *relation*. [...]
>
> A *proposition* is the presentation of two names under a relation. A *judgment* is the sentence of the mind upon a proposition, true, false, more or less probable [...]
>
> The distinction of *subject* and *predicate* is the distinction between the *notion in relation* and the *notion to which it is in relation*.
>
> Every relation has its counter-relation. Or *converse* relation: thus is X be in the relation A to Y, Y is therefore in some relation B to X: and A and B are converse relations, and the propositions are converse propositions. Every proposition has its converse, of meaning identical with itself. [...]
>
> When X has a relation (A) to that which has a relation (B) to Y, X has to Y a *combined* relation: the *combinants* are A and B.[40]
>
> The perception of the validity of a syllogism is the perception of the combination of two relations into one.[41]

Daniel Merrill summarizes in a quite clear and vivid way De Morgan's attitude towards syllogism and relations:

> De Morgan's analysis of the syllogism is shaped by his logic of relations, for he construes the theory of the syllogism in terms of the logic of relations. Conversely, his logic of relations is developed with the theory of the syllogism in mind. It is developed only to the point where the theory of the syllogism can be construed within it.[42]

In *First Notions*, De Morgan introduces the argument "called *a fortiori*" proposing an inference that is an instance of composition of relations:

---

[40] De Morgan (1966: 119).   [41] De Morgan (1966: 130).   [42] Merrill (1990: 149).

(1) All the As make up part (and part only) of the Xs,

(2) All the Xs make up part (and part only) of the Bs,

(3) Therefore, all the As make up part of part (only) of the Bs.

We might regard here 'part of part' as a first instance of composition of relations, a concept that De Morgan will pursue further in his fourth essay on syllogism and the logic of relations (1859).[43]

The path that led De Morgan to develop an embryonic theory of relations, putting together relations and quantification has been carefully reconstructed by Merrill (1990), Panteki (1991), and Sánchez Valencia (2004). These authors have developed a detailed account of De Morgan's thought about syllogism and the logic of relations, and it is to them that we refer the reader interested in the issue. Even though, as Tarski (1941: 73) remarked, De Morgan "cannot be regarded as the creator of the modern theory of relations, since he did not possess an adequate apparatus for treating the subject," he "directed his attention to the general concept of relations and fully recognized its signifi-cance." Moreover, he made the first substantial step toward the construction of a theory that was later developed by Charles Sanders Peirce, the authentic "creator of the theory of relations" and Ernst Schröder.[44]

De Morgan was one of the last logicians in the tradition of Western logic to attempt a transformation of a mathematical (geometrical) demonstration into a chain of syllogisms and he clearly recognized that the traditional (Aristotelian) syllogism was unfit for the job. He maintained, however, that properly improved, the syllogism can still be a basic tool for demonstrating mathematical theorems, the improvement consisting in considering the syllo-gism "under the aspect of combination of relations."[45] Thus, his peculiar views on syllogism led De Morgan to discover a new field of logic: the logic of relations.

---

[43] This remark is due to Panteki (1991, II: 431).   [44] Tarski (1941: 73).

[45] De Morgan (1966: 241); Panteki (1991: 423).

Conclusion

Our history of the attempts to syllogize geometry stops with De Morgan. We have not been able to find other significant attempts to syllogize geometrical theorems after him and his successful attempts at a logic of relations makes De Morgan a natural stopping point for our story. Of course, one should be careful not to think that the work of Frege, Russell, Peirce, Schröder, Peano, among others, immediately brought about the downfall of the "syllogizing" ideology. Logic books in the second half of the nineteenth century (with later editions published well into the twentieth century) still debated the issue. For instance, in Sigwart's *Logik* (first edition 1873; second edition 1888; English translation of the second edition 1895) we find both the claim that all mathematical proofs can be reframed as chains of syllogisms in *Barbara* as well as the claim that this reframing is quite artificial and unsatisfactory when it comes to capturing certain types of relational inferences. On the first claim we read:

> Mathematics, which throughout makes use of syllogism and owes its scientific certainty to this form, is often cited as an answer to all attacks upon the theory of the syllogism. This is quite justifiable so far as showing that all mathematical propositions, with the exception of axioms and definitions, are proved by syllogisms, or at any rate on the principles by which the syllogistic logic forms are determined.[1]

Sigwart qualified his judgment by pointing out that one should not overlook the difference between mathematical propositions and the analytic judgments of scholastic logic. The premises of geometrical reasonings are not always judgments of subsumption and thus although the syllogisms of geometry "may appear to be framed in the scholastic form *Barbara*, they are not so in reality."[2] In addition, he pointed out that several types of intuitive inferences are based on various relations, such as equality, inequality, greater than, and others. Discussing an example found in Ueberweg, Sigwart stated:

---

[1] Sigwart (1895: 362–3).   [2] Sigwart (1895: 363).

Syllogistic Logic and Mathematical Proof. Paolo Mancosu and Massimo Mugnai, Oxford University Press.
© Paolo Mancosu and Massimo Mugnai 2023. DOI: 10.1093/oso/9780198876922.003.0010

The inference does not proceed by means of such subordination at all, but solely by means of relations which form no part of the concept triangle. Where two or more triangles are given, having proportional sides, it follows that the other relation – equality of angles – will also be present; and because equality of angles involves the similarity of the triangles it follows that the relation of similarity is given together with the relation of proportionality of sides. It is only by a crude inaccuracy of expression that these propositions can take the form of a proposition concerning "all triangles" of a particular construction.[3]

The possibility of stating general geometrical laws, however, allows for a recasting of mathematics syllogistically:

It is significant that the guiding principle of mathematical inferences is the proposition that two magnitudes which are equal to a third are equal to each other; that is, it is a proposition concerning necessary connections between relations. Equally significant is it that mathematical reasoning frequently proceeds by substituting one magnitude for another to which is equal. Processes such as these find no place in the ordinary forms of the syllogism, though by means of these general laws they can always be exhibited strictly syllogistically.[4]

From even a cursory probing of the literature in logic from the end of the nineteenth century to the 1920s, it is evident to us that one could come up with more sources to chronicle this tension between relational reasoning and syllogisms.[5] What is of interest in this literature, but it is also what makes

---

[3] Sigwart (1895: 364).   [4] Sigwart (1895: 364).

[5] Bradley's *The Principles of Logic* (1883, 1922) is quite interesting in this regard, for he makes the validity of certain relational inferences the cornerstone of his approach to logic and judges syllogism insufficient to account for them. In Book II, part I, chapter I, §4 he wrote:

Except in the interest of a preconceived theory, I think that these statements, at least so far, will not be denied. But I can hardly hope that the examples of reasoning I am about to produce will all escape unchallenged. Yet I shall not defend them, for I do not know how. They are palpable inferences, and the fact that they are so is much stronger than any theory of logic. (i) A is to the right of B, B is to the right of C, therefore A is to the right of C. (ii) A is due north of B, B due west of C, therefore A is north-west of C. (iii) A is equal to (greater or less than) B, B is equal to (greater or less than) C, therefore & c. (iv) A is in tune with B, and B with C, therefore A with C. (v) A is prior to (after, simultaneous with) B, B to C, therefore A to C. (vi) Heat lengthens the pendulum, what lengthens the pendulum, makes it go slower, therefore heat makes it go slower. (vii) Charles I was a king; he was beheaded, and so a king may be beheaded. (viii) Man is mortal, John is man, therefore John is mortal. We shall go from these facts to ask how far certain theories square with them.   (1883: 226; 1922: 246)

Bradley claims that syllogism is insufficient to account for these inferences: "It is evident that the syllogism can not be saved or can only be saved in such a way as to be syllogism no longer" (1883: 231; 1922: 250).

such a reconstruction less appealing as a general project, is the lack of attention to the recent developments in mathematical logic.[6]

This is, of course, not true of the great innovators of late nineteenth-century logic but even there it would be useful to have a more detailed study—something which goes beyond the framework of our study—of the positions of Frege, Peirce, Schröder, and Peano, to name only the foremost authorities in mathematical logic during that period, with respect to the issue we are after. Peirce for instance, displays an interesting development as witnessed by the following quotes.[7]

Now in order to show the importance of a study of logic, even of the despised syllogism, I will mention that logicians have found such extreme difficulty in reducing mathematical demonstrations to syllogistic form, that some have boldly pronounced it impossible, and on that impossibility have founded a peculiar philosophy of mathematics, and of space and time with which it deals. But a further study of syllogism has led to the discovery of these new forms, by which mathematical demonstrations *can* be reduced to syllogism, . . .[8]

But by 1898 he had reached the conclusion that syllogistic logic was insufficient for mathematics:

But to return to the state of my logical studies in 1867, various facts proved to me beyond a doubt that my scheme of formal logic was still incomplete. For one thing, I found it quite impossible to represent in syllogisms any course of reasoning in geometry or even any reasoning in algebra except in Boole's logical algebra.[9]

---

[6] A further complicating factor is the tendency, evident in Poincaré and others, starting from the end of the nineteenth century to use "syllogism" to include any logical inference. We will not delve into this aspect of the topic.

[7] We are grateful to Sun-Joo Shin who brought to our attention the two passages in Peirce.

[8] Peirce, *Lowell Lectures*, 1866, in Peirce (1982: 385–6).

[9] Peirce continues:

Moreover, I had found that Boole's algebra required enlargement to enable it to represent the ordinary syllogisms of the third figure; and though I had invented such an enlargement, it was evidently of a makeshift character, and there must be some other method springing out of the idea of the algebra itself. Besides, Boole's algebra suggested strongly its own imperfection. Putting these ideas together I discovered the logic of relatives. I was not the first discoverer; but I thought I was, and had complemented Boole's algebra so far as to render it adequate to all reasoning about dyadic relations, before Professor De Morgan sent me his epoch-making memoir ["On the Syllogism IV, and on the Logic of Relations," *Cambridge Philosophical Transactions*, vol. 10: 331–58] in which he attacked the logic of relatives by another method in harmony with his own logical system. But the immense

Of course, the full development of the theory of relations played an important part in this change of heart. In *The Principles of Mathematics*, Russell says:

> There was, until very lately, a specific difficulty in the principles of mathematics. It seemed plain that mathematics consists of deductions, and yet the orthodox accounts of deduction were largely or wholly inapplicable to existing mathematics. Not only the Aristotelian syllogistic theory, but also the modern doctrines of Symbolic Logic, were either theoretically inadequate to mathematical reasoning, or at any rate required such artificial forms of statement that they could not be practically applied.[10]

Indeed, according to Russell, even the development of a theory of relations was not sufficient to grasp its relevance to an improved theory of deductive logic as it was originally so shrouded in an unclear formalism and surrounded by a philosophical fog that it took time to finally see how much the theory of relations had changed the situation. Here is Russell's explanation of why it took so long to finally see things aright.

> A careful analysis of mathematical reasoning shows (as we shall find in the course of the present work) that types of relations are the true subject-matter discussed, however a bad phraseology may disguise this fact; hence the logic of relations has a more immediate bearing on mathematics than that of classes or propositions, and any theoretically correct and adequate expression of mathematical truths is only possible by its means. Peirce and Schröder have realized the great importance of the subject, but unfortunately their methods, being based, not on Peano, but on the older Symbolic Logic derived (with modifications) from Boole, are so cumbrous and difficult that most of the applications which ought to be made are practically not feasible. In addition to the defects of the old Symbolic Logic, their method suffers technically (whether philosophically or not I do not at present discuss) from the fact that they regard a relation essentially as a class of couples, thus requiring elaborate formulae of summation for dealing with single relations. This view is derived, I think, probably unconsciously, from a philosophical

superiority of the Boolian method was apparent enough, and I shall never forget all there was of manliness and pathos in De Morgan's face when I pointed it out to him in 1870. I wondered whether when I was in my last days some young man would come and point out to me how much of my work must be superseded, and whether I should be able to take it with the same genuine candor... (Peirce 1931–58, Vol. 4: 4)

[10] Russell (1903: 4).

error: it has always been customary to suppose relational propositions less ultimate than class-propositions (or subject-predicate propositions, with which class-propositions are habitually confounded), and this has led to a desire to treat relations as a kind of class. However this may be, it was certainly from the opposite philosophical belief, which I derived from my friend Mr G. E. Moore, that I was led to a different formal treatment of relations. This treatment, whether more philosophically correct or not, is certainly far more convenient and far more powerful as an engine of discovery in actual mathematics.[11]

And, to repeat a quote we gave in the section on Kant:

It was still possible to hold, as Kant did, that no great advance had been made since Aristotle, and that none, therefore, was likely to occur in the future. The syllogism still remained the one type of formally correct reasoning; and the syllogism was certainly inadequate for mathematics. But now, thanks mainly to the mathematical logicians, formal logic is enriched by several forms of reasoning not reducible to the syllogism, and by means of these all mathematics can be, and large parts of mathematics actually have been, developed strictly according to the rules.[12]

So, was Russell the end of the story? Not quite. One should not make the mistake of thinking that as soon as a proper theory of quantification and relational inferences had been developed by the mathematical logicians that this immediately resulted in a general recognition of the insufficiency of syllogisms for capturing the logic of mathematical reasoning.

The most intriguing counterexample to such a thesis is given by Moritz Schlick's *Allgemeine Erkenntnislehre* (1918, 1925). Even in its second edition published in 1925, Schlick, while fully aware of Russell's work in logic, continued to defend the idea that all scientific knowledge, and mathematics in particular, can be presented in syllogisms in *Barbara*:[13]

That the interconnection of truths in rigorous scientific systems can indeed be represented by means of this form of inference is shown by any inquiry into such systems. The only reason an investigation is needed to confirm this

---

[11] Russell (1903 [1939]: 24).   [12] Russell (1903 [1939]: 457).

[13] One can trace back Schlick's position on the power of syllogistic logic, with further details on its sources, to the lecture course "Grundzüge der Erkenntnistheorie und Logik" delivered in Rostock in 1911–12. See Stelling (2019: 368–659) and Lemke and Naujoks (2019).

fact is that scientific deductions are almost always presented in abbreviated rather than pure syllogistic form. In particular, the minor premisses for the most part are not stated separately, since they can easily be gathered from the sense, and the trained thinker usually hurries by them. The prime example of a tightly interconnected system of scientific truths that comes naturally to mind is mathematics. Here individual propositions are linked together by those processes we call proofs and calculations. These are nothing other than sequences of syllogisms in the mood *Barbara*. In principle, all demonstrations proceed according to the same schema, in the form illustrated by the following example:

Every right triangle is endowed with such and such properties.
The figure ABC is a right triangle.

ABC is endowed with such and such properties.

The major premiss states a general rule (proved, in turn, from still more general propositions), under which the syllogism subsumes the particular subject of the minor premiss. The correctness of the latter, however, rests either directly on definition (or, in the language of geometry, construction) or indirectly on a proof that takes the proposition back to the fundamental definitions (axioms) of geometry.[14]

Schlick took issue with Sigwart. While conceding that Sigwart was right that mathematical inferences in practice do not look like simple syllogisms, he objected to Sigwart's claim that the major premises of such arguments could not be recast in the form of subsumptions.

Geometrical demonstrations are of this kind. Sigwart is right when he objects to taking as the paradigm of mathematical inference a simple syllogism such as: A parallelogram is a quadrilateral; a square is a parallelogram; hence a square is a quadrilateral. He is wrong, however, when he concludes further that the major premises of geometrical inferences cannot in general be conceived of as subsumption judgments and that they only seem to have the form of the mood *Barbara*. Specifically, his view is that geometry does not deal only with the subsumption relationship of concepts. Rather, it "always goes beyond mere conceptual judgments"; it derives its propositions "with the aid of law-like relations taken from somewhere or other" (this

----

[14] Schlick (1985: 104–5).

"somewhere or other" must obviously be intuition), relations not contained in the definitions.[15]

Schlick appealed to his view of formal systems as implicit definition to counter Sigwart's position.

> To counter this view, we need only recall some of our earlier discussion (Part I, § 7). We saw that a modern rigorous system of geometry uses just those relations that are contained in the definitions. Indeed, the definition of the basic concepts of the system takes place precisely through these relations, and that is why the laws governing the relations may be represented as subsumption relationships of concepts, and conversely. Sigwart, still the captive of older views on the nature of mathematical thought, overlooked this point when he insisted that mathematical reduction proceeds not on the basis of subsumption relationships of concepts but on the basis of relationships of relations. But from a purely logical and mathematical standpoint, the two are one and the same, since strict, pure concepts are simply nodal points of relations. What is true of geometry is true in similar fashion of arithmetic and algebra.[16]

In this way he could conclude his defense of syllogism in all areas of knowledge:

> It thus becomes clear that the most rigorous systems of knowledge can be rendered by means of the mood *Barbara*. From the standpoint of pure logic, there is no distinction between the rigorous inferences of any arbitrary science and those of mathematics; for in treating inferences, we consider only the relationship of concepts to one another, without regard to the various intuitive objects that are designated by means of these concepts. Hence all truths that have precise logical interconnection (that is, that admit of being deduced from one another) can be represented as far as their mutual linkage is concerned by means of syllogisms, specifically in the mood *Barbara*.[17]

Against the possible objection that the work of recent logicians (including Russell) had shown the inadequacy of syllogisms for capturing the logic of

---

[15] Schlick (1985: 105).   [16] Schlick (1985: 105–6).   [17] Schlick (1985: 106–7).

mathematical inference he made a distinction between the forms of inference used in practice versus the reconstruction 'in principle':

> The Aristotelian theory of inference needs no modification or extension in order to be applicable to modern science. What is necessary is only that the theory of concepts be deepened, and this has already taken place, as indicated in part of the preceding discussion. Modern logic, in the form developed by Bertrand Russell, for example, no doubt offers a much more useful set of inference procedures than the syllogistic. Beyond this, however, all the arguments advanced against the dominion of the syllogism prove only that the actual thought of man does not proceed in regular syllogisms – and this is an undeniable psychological fact. But they fail to refute the thesis that presentation of an absolutely rigorous system of truth, so far as the presentation is meant to be absolutely exact and complete, can always take place in syllogistic form. And it is only this thesis that must be maintained here. It is quite obvious, for example, that the actual discovery of geometrical truths needs by no means follow the pattern of *Barbara*. The process of discovery may involve the use, say, of negative judgments (as in so-called indirect proof). But this does not affect the inner ties which necessarily connect the individual propositions and on which our examination is centered.[18]

Schlick's position highlights the "ideological" power of the syllogizing picture, which could almost be described as an "influential metaphysics."

We now need to conclude with the inevitable question: Why did it take so long to realize that syllogisms are insufficient to account for the logic of mathematics?

In the Western tradition, since antiquity, logic and mathematics developed as two quite independent disciplines. Aristotle, in his logical works occasionally refers to some mathematical (geometrical) demonstrations and, on the side of mathematics, authors who developed commentaries on Euclid's work made rare appeals to syllogisms or to rules belonging to the logical canon of the time. During the Middle Ages, logic and mathematics were taught in different sections of the academic curriculum: logic in the so-called *trivium*, mathematics (arithmetic, geometry) in the *quadrivium*.

As we have seen, in 1566 Herlinus and Dasypodius made, for the first time, a systematic attempt to give syllogistic form to the demonstrations of the first six books of the *Elements*. Later on, several centuries since Galen, in the first

---

[18] Schlick (1985: 107).

half of the seventeenth century, Joachim Jungius and his pupil Johannes Vagetius emphasized that for performing mathematical demonstrations it was necessary to include, among the tools of traditional logic, inferences based on relations. Jungius and, in particular, Vagetius were well aware that certain inferences employing relations cannot be reduced to syllogistic form and that they are indispensable for proving many geometrical theorems.

Herlinus and Dasypodius may be considered the representatives of a kind of movement from logic toward mathematics (geometry), insofar as they attempted to show that the traditional logical apparatus mainly based on the Aristotelian syllogism was sufficient to account for the demonstrations included in the *Elements*. In the seventeenth century, however, another movement arose in the opposite direction, as it were, from mathematics toward logic, attempting to reduce all logical inferences to a calculus analogous to those usually performed in algebra. Leibniz and Jacob Bernoulli are the outstanding representatives of this second movement.[19] Leibniz, in his *Dissertation on combinatorial art*, enthusiastically adhered to Hobbes' claim according to which *thinking is computing*, a claim quite revolutionary at the time, that could be assumed as *motto* of the project for the constitution of the Leibnizian 'characteristic art.' Today we know that Leibniz worked out a logical calculus analogous to that later discovered by George Boole, but he never published his logical essays, a conspicuous part of which was edited only at the beginning of the twentieth century by Louis Couturat.[20]

Had Leibniz published his essays on logical calculus, the history of logic in Western culture would probably have been quite different. At any rate, it was only with the sixteenth and the seventeenth centuries that logic and mathematics began for the first time, as it were, 'to speak to one another.' And it was during the seventeenth century that Jungius and Leibniz raised again the problem of the logical form of relational propositions in a setting of traditional logic mainly inspired by the Aristotelian doctrine.

During the Middle Ages, an intensively discussed topic among logicians was the nature and behavior of the so called 'consequences.' A *consequence* was usually defined as an argument having a non-syllogistic form, composed of an antecedent and a consequent joined by expressions like 'therefore,' 'thus,' 'if,'

---

[19] Bernoulli and Bernoulli (1685).

[20] Cf., Couturat (1903). Before Couturat, some texts on logic were edited in Erdmann (see Leibniz 1840) and in vol. 7 of Leibniz (1875–90). On Leibniz and Boolean algebra, see Castañeda (1976), Lenzen (1984, 2004), and Malink and Vasudevan (2016, 2017). On the reception of the logical writings by Leibniz in the nineteenth century, see Peckhaus (1997).

'because,' and so on. This is, for example, the definition of a consequence given by Buridan:

> To this end, I say that propositions are divided into subject and predicate and compound propositions. Now a consequence is a compound proposition, for it is constituted from several propositions conjoined by the expression 'if' or the expression 'therefore' or something equivalent. For these expressions mean that of propositions conjoined by them one follows from the other; and they differ in that the expression 'if' means that the proposition imme- diately following it is the antecedent and the other the consequent, but the expression 'therefore' means the converse.[21]

According to this definition, conditionals of the general form 'If p, then q' and inferences of the form 'p, therefore q' are both *consequences* (*consequentiae*). Medieval logicians wrote several treatises devoted to the rules governing consequences. To give an idea of what these rules looked like, let us quote some of them from Walter Burley's *Longer Treatise*:

> The second principal rule is that what follows from the consequent, follows from the antecedent. And there is another rule almost identical with this, which reads: anything antecedent of the antecedent is an antecedent of the consequent.
>
> [...] Anything following from the antecedent and the consequent, follows from the antecedent *per se*.
>
> [...] Anything following from the consequent, follows from the antecedent joined with the same consequent.
>
> [...] A third rule that follows is that in every valid consequence, the opposite of the consequent is repugnant to the antecedent.[22]

These rules, together with many others (among which *modus ponens* and *modus tollens*) belonged to a large list of rules that, with some negligible variations, constituted the core of the theory exposed in the treatises on consequences. What we want to stress and what we find puzzling is that, as far as we know, in the entire logical literature of the scholastic period nothing similar exists for the case of rules governing relations and relational sentences.

---

[21] Buridan (2015: 66).   [22] Burleigh (1955: 62–3).

Of course, some relational inferences were well known and discussed by medieval logicians; an example is offered by the handbook of logic of William of Lucca, a twelfth-century author inspired by the doctrines of Abelard:

> From the relatives, based on the rule that relatives mutually demand each other, we get this argument: "there is paternity or father, therefore there is filiation or son." Also, "this is the father of that, therefore he is not the son of that same," by virtue of the rule that relatives cannot agree under the same with respect to the same thing. Also, "Socrates is the father of Plato, therefore Plato is the son of Socrates," by virtue of the rule that relatives refer to each other.[23]

As can be seen, the rules recalled in the text faithfully reproduce what Aristotle stated in the chapter on relatives (*ad aliud*) in the *Categories*. William of Lucca limits himself to these sporadic remarks and does not attempt to develop the topic of relational inferences further. Other medieval logicians behave similarly when dealing with the topic of relations. Buridan, as Stephen Read, Juan Manuel Campos Benítez, and, recently, Boaz Faraday Schuman have shown, developed a treatment of inferences containing multiple generality and relations that is quite refined and a prelude to a logic of relations in the proper sense.[24] Boaz Faraday Schuman has highlighted with great precision and clarity the complexity of the investigations of Buridan and his school regarding multiple generality. The logical domain in which Buridan moves, however, is that of the traditional Aristotelian square of oppositions: the propositions he deals with are propositions such as 'All some man's horses run,' 'Some man's horses do not run,' and the inferences are, for instance, from a proposition like 'Of some man, every donkey runs' to its contradictory, etc. Buridan expands the Aristotelian square, but he seems not to be interested in working out a general outline of a theory of relational inferences.

Relations and relational sentences abound in everyday language and in theological and philosophical discourse as well, but the medieval logicians did not focus on them and directed their attention towards other topics like the theory of consequences or even the so-called 'supposition theory,' which implied sophisticated reflections on quantification associated with a complex notion of reference.

---

[23] William, Bishop of Lucca (1975: 170).
[24] Campos Benítez (2014); Read (2012), (2016); Schuman (2022). We are grateful to Boaz Faraday Schuman for having made the text of his paper available to us before it was published.

As a matter of fact, logicians in the Western tradition began to be interested in relations and relational sentences as soon as logic, with Jungius, Leibniz, Vagetius, and Lambert made a closer contact with mathematics. Later on, in the second half of the nineteenth century, a first, embryonic form of a logic of relations was developed by Augustus De Morgan, a mathematician strongly interested in logic. This seems to be something more than pure chance: relations are extremely important in mathematics and so are relational sentences.

Even though we assume that it has been the interaction with mathematics, promoted by the re-discovery of Euclid, to awaken a first interest of the logicians for the logic of relations, the question still remains open why before the seventeenth century, in antiquity and in the Middle Ages, independently of any contact with mathematics, philosophers and logicians were oblivious to the importance that relation and relational sentences have in ordinary discourse and therefore in the arguments usually displayed in everyday life.

The nature of relations is quite elusive: whereas we have the impression of being well acquainted with properties as being red, or warm or heavy that we attribute to the objects of our everyday experience, we do not perceive relations in the same way as we perceive something as red, or warm or heavy. If John seats on my left side, I may see John sitting and my left arm, but I cannot see the relation of being situated on the left of myself in the same way as I see the color of John's shirt. This is probably a reason why for centuries philosophers have argued about the ontology of relations.

The view of the world as composed of individuals endowed with their intrinsic properties—an Aristotelian view of the world, as it were—seems to be quite natural and in agreement with the 'spontaneous' and immediate experience of what surrounds us in our everyday life. Typically, when Aristotle discusses the category of 'being toward something' (in Greek 'pròs ti', in Latin *ad aliud*: traditionally interpreted as an expression denoting relations), he treated, for the most part, *relational properties* instead of *relations*. This giving prominence to properties agrees with Aristotle's idea of how a scientific investigation has to be conducted. Science, as devised by Aristotle was mainly based on the notion of *essence* and aimed to develop classifications, distributing things into appropriate classes: relations, in Aristotle's view of science, played a subordinate role compared with essences. Thus, the preeminent role played by Aristotelian philosophy until the sixteenth century, may explain the lack of interest of logicians and philosophers for a logic of relations. Given the central position that the Aristotelian syllogism occupied in traditional logic, it is not difficult to imagine that people simply assumed as something obvious that relations and relational sentences could be easily integrated into the syllogistic machinery.

Relations and relational sentences, however, play an outstanding role in Euclid's *Elements* as well, and the reason why logicians in antiquity, with the exception of Galen, did not notice their importance seems to reside mainly in the separation of the two disciplines. Logicians in antiquity and in the Middle Ages showed, by and large, no particular interest for mathematical demonstrations; and the mathematicians of the same periods did not care about the traditional logical doctrine of Aristotelian origins. Obviously, mathematicians were well acquainted with relations and relational sentences, but to demonstrate theorems they took recourse to the techniques employed by Euclid and other eminent mathematicians, not to the Aristotelian *Prior Analytics*. Even today, to prove theorems mathematicians do not make explicit use of first order logic as it is displayed in contemporary logic handbooks.

The long (and subtle) persistence of Aristotle's authority in the field of logical theory, together with a slow process of 'mathematization' of logic, explain why so much time was needed before logicians could become aware of the importance of relations with their logical properties. Relations began to gain the front stage only when mathematics entered the play and physics ceased to be a merely *qualitative* science (whether this begins with Galileo or earlier in the Renaissance is not essential here). In the same period, as we have seen, under the effect of a violent reaction against the sophistries of the scholastic teaching, together with the re-discovery of important works of antiquity, such as Proclus's Commentary on the first book of Euclid's *Elements*, Euclid's *Elements*, not Aristotle's *Analytics*, became the standard of rigorous demonstration. Thinking was equated to computing and François Viète's algebraic investigations fueled the expectations for the constitution of a symbolism apt to develop a logical calculus. To investigate the logical rules to which relations are supposed to obey, taking recourse to the only resources of natural language, is more difficult than doing the same in the case, for instance, of conditionals. Thus, it is no wonder that Leibniz, Caramuel, and Lambert, for example, in their logical essays, as we have seen, attempted to avail themselves of special symbols to express sentences involving relations. From the seventeenth century onwards, mathematics and mathematical symbolism began to take possession, even though at a slow pace, of the field of logic, this process culminating with the works of George Boole (*The Mathematical Analysis of Logic*, 1847) and Gottlob Frege (*Begriffsschrift*, 1879).

It was not mere coincidence that De Morgan, before publishing his seminal essay on relations, edited a paper on the calculus of functions (1836), where he pointed out some operations with functions analogous to those that hold for relations (as, for instance, the composition of functions): clearly his activity as a mathematician had strongly influenced his ideas about the syllogism.[25] This is clearly witnessed by the following passages, extracted, respectively, from the third and the fourth essay on syllogism (1858):

> The syllogism of each of the opposed, or rather confronted, wholes of mathematical and metaphysical thought, is of second intention, and deals with the notions so called. It is, in fact, *combinations of relations*: the act of mind by which we see that the *A* of (the *B* of *Z*), or the (*A* of *B*) of *Z*, is thinkable under one relation. Here the compound relation, or combined relation, may be represented by *AB*, but by no one except a mathematician who is used to the *functional* symbol, and to the idea of $\varphi\psi(xy)$ in its distinction between the mode of composition of *x*, *y*, and that of φ, ψ.[26]

> It is to *algebra* that we must look for the most habitual use of the logical forms. [...] [S]o soon as the syllogism is considered under the aspect of combination of relations, it becomes clear that there is more of syllogism, and more of its variety, in algebra than in any other subject whatever, though the matter of the relations – pure quantity – is itself of small variety. And here the general idea of relation emerges, and for the first time in the history of knowledge, the notions of relation and *relation of relation* are symbolized.[27]

In 1868, in the London literary magazine *Athenaeum* which we have used as the epigraph of our book, De Morgan complained about the state of separation of logic and mathematics: "we know that mathematicians care no more for logic than logicians for mathematics," but it was in great part thanks to him that the situation soon began to change.

---

[25] In 1836 De Morgan published in the *Encyclopedia Metropolitana* the treatise "Calculus of functions."

[26] De Morgan (1966: 107).

[27] De Morgan (1966: 241). Cf., Panteki (1991: 203):

> As far as the methodology for the solution of functional equations was concerned, little improvement could be affected [...] upon the methods of Laplace, Babbage and Herschel [...] However, the study of foundational issues, such as the properties of inverse, convertible, periodic and other functions [...], formed the most challenging ground for De Morgan to exert for the first time his critical reasoning and his tendency to delve into the 'nature of things', deduce all possible relations between symbols and vindicate the view – the germ of which were to be found in the work of his English (and French) predecessors [...] that symbolic relations "are worth looking at as modes of invention" [...].

References

Albert of Saxony [Albert von Sachsen] (2010). *Logik*, (Lateinisch-Deutsch), Übersetzt, mit einer Einleitung und Anmerkungen, herausgegeben von Harald Berger, Hamburg: Meiner Verlag.

Albertus Magnus (2003). *The Commentary of Albertus Magnus on Book I of Euclid's Elements of Geometry*, ed. Anthony lo Bello, Boston/Leiden: Brill Academic Publishers.

Alexander of Aphrodisias (2006). *On Aristotle's Prior Analytics I.23-31*, trans. Ian Mueller, Ithaca: Cornell University Press.

Anderson, Lanier (2005). 'The Wolffian Paradigm and its Discontents: Kant's Containment Definition of Analyticity in Historical Context,' *Archiv für Geschichte der Philosophie* 87: 22-74.

Anderson, Lanier (2015). *The Poverty of Conceptual Truth: Kant's Analytic/Synthetic Distinction and the Limits of Metaphysics*, Oxford: Oxford University Press.

Antognazza, Maria Rosa (2009). *Leibniz: An Intellectual Biography*, Oxford: Oxford University Press.

Aristotle (1984). *The Complete Works*, ed. Jonathan Barnes, Princeton: Princeton University Press.

Aristotle (1989). *Prior Analytics*, trans., with introduction, notes, and commentary, Robin Smith, Indianapolis: Hackett.

Arnauld, Antoine and Nicole, Pierre (1682). *Logica, sive ars cogitandi*, Lugduni: apud J. Gaal.

Arnauld Antoine and Nicole Pierre (1981). *La Logique ou l'Art de Penser*, édition critique par Pierre Clair et Francois Girbal, Seconde édition revue, Paris: Vrin.

Arndt, Hans Werner (1965). Introduction to Christian Wolff, *Vernünftige Gedanken von den Kräften des menschlichen Verstandes* (German Logic), in C. Wolff, *Gesammelte Werke*, I.Abt., Band I, Hildesheim: Georg Olms: 7-102.

Arndt, Hans Werner (1971). *Methodo Scientifica Pertractatam, Mos geometricus und Kalkülbegriff in der philosophischen Theorienbildung des 17. und 18. Jahrhunderts*, Berlin: de Gruyter.

Ashworth, Jennifer E. (1967). 'Joachim Jungius (1587-1657) and the Logic of Relations,' *Archiv für Geschichte der Philosophie* 49: 78-85.

Ashworth, Jennifer E. (1974). *Language and Logic in the Post-Medieval Period*, Dordrecht/Boston: D. Reidel Publishing Company.

Auriol, Peter (1596-1605). *Commentariorum in primum [-quartum] librum sententiarum pars prima*. Romæ: ex Typographia Vaticana.

Avicenna (1971). *Treatise on Logic: Part One of Danesh-name Alai; A Concise Philosophical Encyclopaedia and Autobiography*, ed. and trans. F. Zabeeh, The Hague: Nijhoff.

Avigad, Jeremy, Dean, Edward, and Mumma, John (2009). 'A Formal System for Euclid's Elements,' *The Review of Symbolic Logic* 2(4): 700-68.

Barnes, Jonathan (2007). *Truth, etc. Six Lectures on Ancient Logic*, Oxford: Clarendon Press.

Barnes, Jonathan (2012). 'Logical Form and Logical Matter,' in Barnes, Jonathan, *Logical Matters. Essays in Ancient Philosophy* II, Oxford: Oxford University Press: 43-146.

Barrow, Isaac (1664). *Lectiones Mathematicae*, London: Typis J. Playford, pro Georgio Wells.

Basso, Paola (2004). *Il secolo geometrico. La questione del metodo matematico in filosofia da Spinoza a Kant*, Firenze: Le Lettere.

Beck, Lewis White (1955-6). 'Can Synthetic Be Made Analytic?,' *Kant-Studien* 47: 168-81.

Beck, Lewis White (1956). 'Kant's Theory of Definition,' *Philosophical Review* 65: 179-91.

Beck, Lewis White (1965). *Studies in the Philosophy of Kant*, Indianapolis: Bobbs-Merrill.

Beeckman, Isaac (1942). *Journal tenu par Isaac Beeckman de 1604 à 1634*, introduction and notes by C. de Waard, Tome Deuxième 1619-27, The Hague: Martinus Nijhoff.

Beeson, Michael (2015). 'A Constructive Version of Tarski's Geometry,' *Annals of Pure and Applied Logic* 166 (11): 1199-273.

Beeson, Michael (2022). 'Euclid after Computer Proof-Checking,' *American Mathematical Monthly* 129: 623-46.

Beeson, Michael, Narboux, Julien, and Wiedijk, Freek (2019). 'Proof-Checking Euclid,' *Annals of Mathematics and Artificial Intelligence* 85: 213-57.

Bernoulli, Jacques (1969). *Die Werke von Jakob Bernoulli*, Band 1, Basel: Birkhäuser.

Bernoulli, Jacques and Bernoulli, Jean (1685). *Parallelismus ratiocinii logici et algebraici*. Reprinted in *Jacobi Bernoulli, Basileensis, opera* vol. 1 (Geneva 1744): 211-24 and Bernoulli (1969: 263-74).

Bertato, Fabio Maia (2014). 'Sobre as formalizações silogísticas dos Elementos, efetuadas por Herlinus, Dasypodius, Clavius e Hérigone,' in S. Nobre, F. Bertato, and L. Saraiva (eds.), *Anais/Actas do 6o Encontro Luso-Brasileiro de História da Matemática*, Natal: Sociedade Brasileira de História Matemática: 927-62.

Beth, Evert (1956-7). 'Über Lockes "Allgemeines Dreieck,"' *Kant-Studien* 48: 361-80.

Beth, Evert (1957). *La Crise de la Raison et la Logique*, Paris: Gauthier-Villars.

Bird, Otto (1964). *Syllogistic and Its Extensions*, Englewood Cliffs: Prentice-Hall.

Bochenski, Joseph Maria (1968). *Ancient Formal Logic*, Amsterdam: North-Holland Publishing House (first edn. 1951).

Bolzano, Bernard (1810). *Beyträge zu einer begründeteren Darstellung der Mathematik Erste Lieferung*, Prague, XVI + 152 pp. Trans. as *Contributions to a Better-Grounded Presentation of Mathematics* in Russ (2004): 83-137.

Bolzano, Bernard (1844). *Wissenschaftslehre*, Sulzbach. Trans. as *Theory of Science*, in Bolzano (2014).

Bolzano, Bernard (1977). *Allgemeine Mathesis*, in *Bernard Bolzano-Gesamtausgabe*, Series 2A, Vol. 5: 15-64, Stuttgart-Bad Cannstatt: Frommann-Holzboog.

Bolzano, Bernard (2014). *Theory of Science*, ed. and trans. George Rolf and Paul Rusnock, Oxford: Oxford University Press.

Boole, George (1847). *The Mathematical Analysis of Logic*, Cambridge: Macmillan, Barclay and Macmillan; London: George Bell.

Boswell, Terry (1991). *Quellenkritische Untersuchungen zum Kantischen Logikhandbuch*, Frankfurt am Main: Peter Lang.

Bozzi, Silvio (2013). 'Review of Saccheri's *Logica demonstrativa*,' *History and Philosophy of Logic* 34: 183-7.

Bradley, Francis Herbert (1883 [1922]). *The Principles of Logic*, London: Oxford University Press.

Brittan, Gordon (1978). *Kant's Theory of Science*, Princeton: Princeton University Press.

Broadie, Alexander (ed.) (2004). *Thomas Reid on Logic, Rhetoric and the Fine Arts*, University Park: Pennsylvania State University Press.

Buridan, John (2001). *Summulae de dialectica*, an annotated translation, with a philosophical introduction by Gyula Klima, New Haven and London: Yale University Press.

Buridan, John (2015). *Treatise on Consequences*, trans. and with an introduction by Stephen Read, editorial introduction by Hubert Hubien, New York: Fordham University Press.

Buridan, John (n.d.). *Quaestiones in Analytica priora*, Liber 1, q. 6 (trans. E. D. Buckner: *The Logic Museum*). http://www.logicmuseum.com/wiki/Authors/Buridan/Quaestiones_in_analytica_priora/Liber_1.

Burleigh [Burley], Walter (1955). *De Puritate Artis Logicae Tractatus Longior*, with a revised edition of the *Tractatus brevior*, ed. Ph. Boehner, St. Bonaventure: Franciscan Institute Publications.

Campo, Mariano (1939). *Cristiano Wolff e il razionalismo precritico*, 2 vols., Milano: Società Editrice Vita e Pensiero (reprint Hildesheim: Georg Olms, 1980).

Campos Benítez, Juan M. (2014). 'The Medieval Octagon of Opposition for Sentences with Quantified Predicates,' *History and Philosophy of Logic* 35 (4): 354–68.

Capozzi, Mirella (1973). 'J. Hintikka e il metodo della matematica,' *Il Pensiero* 18: 232–67.

Capozzi, Mirella (2020). 'Singular Terms and Intuitions in Kant: A Reappraisal,' in Carl Posy and Ofra Rechter (eds.), *Kant's Philosophy of Mathematics. Volume I: The Critical Philosophy and Its Roots*, Cambridge: Cambridge University Press: 103–25.

Caramuel y Lobkowicz, Juan (1654). *Theologia Rationalis*, Frankfurt am Main: J. G. Schönwetter.

Carson, Emily (2009). 'Hintikka on Kant's Mathematical Method,' *Revue Internationale de Philosophie* 250 (4): 435–49.

Cassirer, Ernst (1907). 'Kant und die moderne Mathematik,' *Kant-Studien* 12: 1–40.

Castañeda, Hector-Neri (1976). 'Leibniz's Syllogistico-Propositional Calculus,' *Notre Dame Journal of Formal Logic* 17: 481–500.

Chikurel, Idit (2020). *Salomon Maimon's Theory of Invention: Scientific Genius, Analysis and Euclidean Geometry*, Berlin/Boston: de Gruyter.

Church, Alonzo (1936). 'A Note on the Entscheidungsproblem', *The Journal of Symbolic Logic* 1: 40–1; correction 1: 101–2.

Church, Alonzo (1968). Review of Joachim Jungius, *Logica Hamburgensis*, ed. Rudolf W. Meyer, *The Journal of Symbolic Logic* 33(21): 139.

Ciafardone Raffaele (1978). *L'Illuminismo tedesco: Metodo filosofico e premesse etico-filosofiche (1690–1765)*, Rieti: Il Velino.

Clavius, Christophorus (1574). *Euclidis Elementorum libri XV: accessit XVI de solidorum regularium comparatione*, Rome: V. Accolti.

Clavius, Christophorus (1591). *Commentaria in Euclidis Elementorum Libri XV*, Rome: G. B. Ciotti.

Clavius, Christophorus (1612). *Opera Mathematica, Quinque tomis distributa, ab Auctore nunc denuo correcta, et plurimis locis aucta*, Moguntiae: Reinhard Eltz.

Corcoran, John (1972). 'Completeness of an Ancient Logic,' *Journal of Symbolic Logic* 37: 696–702.

Corcoran, John (1974). 'Aristotle's Natural Deduction System,' in John Corcoran (ed.), *Ancient Logic and Its Modern Interpretations*, Dordrecht: Reidel Publishing Company: 85–131.

Couturat, Louis (1903). *Opuscules et fragments inédits de Leibniz, extraits des manuscrits de la Bibliothèque royale de Hanovre*, Paris: Alcan.

Couturat, Louis (1904). 'La philosophie des mathématiques de Kant,' *Revue de métaphysique et de morale* 1904: 321–83.

Couturat, Louis (1905). *Les Principes des Mathématiques: avec un appendice sur la philosophie des mathématiques de Kant*, Paris: F. Alcan.

Cozzoli, Daniele (2007). 'Alessandro Piccolomini and the Certitude of Mathematics,' *History and Philosophy of Logic*, 28(2): 151–71.

Crapulli, Giovanni (1969). *Mathesis Universalis. Genesi di una idea nel XVI secolo*, Rome: Edizioni dell'Ateneo.

Crivelli, Paolo (2012). 'Aristotle's Logic,' in Christopher Shields (ed.), *The Oxford Handbook of Aristotle*, Oxford: Oxford University Press: 113–49.

Crusius, Christian August (1747). *Weg zur Gewißheit und Zuverläßigkeit der menschlichen Erkenntniß*, Leipzig: Gleditsch.

De Angelis, Enrico (1964). *Il metodo geometrico nella filosofia del Seicento*, Pisa: Ed. Istituto di Filosofia.

De Felice, Federica (2008). *Filosofia e matematica nell'illuminismo tedesco*, Roma: Aracne Editrice.

De Felice, Federica (2016). 'La filosofia di Andreas Rüdiger. Un importante contributo al progresso dell'Aufklärung,' *Paradigmi. Rivista di critica filosofica* 2: 171–84.

De Jong, Willem R. (1995). 'Kant's Analytic Judgments and the Traditional Theory of Concepts,' *Journal of the History of Philosophy* 3(4): 613–41.

De Jong, Willem R. (1998). 'Kant's Theory of Geometrical Reasoning and the Analytic-Synthetic Distinction: On Hintikka's Interpretation of Kant's Philosophy of Mathematics,' *Studies in History and Philosophy of Science* 28(1): 141–66.

De Jong, Willem R. (2010). 'The Analytic-Synthetic Distinction and the Classical Model of Science: Kant, Bolzano and Frege,' *Synthese* 174(2): 237–61.

De Morgan, Augustus (1831 [1943]). *On the Study and Difficulties of Mathematics*, London: Society for the Diffusion of Useful Knowledge. Fourth reprint edition published by The Open Court Publishing Company: La Salle, 1943.

De Morgan, Augustus (1836). 'Calculus of Functions,' in *Encyclopedia Metropolitana*, 2: 305–92.

De Morgan, Augustus (1839). *First Notions of Logic (Preparatory to the Study of Geometry)*, London: Taylor and Walton.

De Morgan, Augustus (1847). *Formal Logic or The Calculus of Inference, Necessary or Probable*, London: Taylor and Walton.

De Morgan, Augustus (1868). Review of *Elementary Geometry*, Part I, Compiled by James M. Wilson, London and Cambridge: MacMillan and Co, (1868), in *Athenaeum*, July 18: 71–3.

De Morgan, Augustus (1966). *On the Syllogism and Other Logical Writings*, ed. with an Introduction by Peter Heath, New Haven: Yale University Press.

De Risi, Vincenzo (2016). 'The Development of Euclidean Axiomatics: The Systems of Principles and the Foundations of Mathematics in Editions of the *Elements* from Antiquity to the Eighteenth Century,' *Archive for History of Exact Sciences* 70: 591–676.

De Vleeschauwer, Hermann Jan (1932). 'La genèse de la méthode mathématique de Wolff. Contribution à l'histoire des idées au XVIIIè siècle,' *Revue Belge de Philologie et d'Histoire* 11(3–4): 651–77.

Dvorák, Petr (2008). 'Relational Logic of Juan Caramuel,' in Dov M. Gabbay and John Woods (eds.), *Handbook of the History of Logic*, vol. 2. Medieval and Renaissance Logic, Amsterdam/Boston/Tokyo: Elsevier/North Holland: 645–65.

Englebretsen, George (1980). 'Singular Terms and the Syllogistic,' *The New Scholasticism* 54: 68–74.

Fabri, Honoré (1646). *Metaphysica Demonstrativa, sive Scientia Rationum Universalium.* Auctore Petro Mousnerio Doctore Medico, Cuncta Excerpta ex Praelectionibus R. P. Honorati Fabri S. J., Lugduni: Sumptibus Johannis Champion.

Folkerts, Menso (1989). *Euclid in Medieval Europe.* The Benjamin catalogue.

Freguglia, Paolo (1988). *Ars Analytica. Matematica e methodus nella seconda metà del Cinquecento,* Busto Arsizio: Bramante.

Freguglia, Paolo (1991). 'Matematica e Logica tra Cinquecento e Seicento,' *Momenti di Storia della Logica dal XVI and XIX Secolo,* Roma: La Goliardica: 21–44 (also in *Epistemologia,* XII, 1989: 115–34).

Freguglia, Paolo (1999). *La geometria tra tradizione e innovazione. Temi e metodi geometrici nell'età della rivoluzione scientifica 1550-1650,* Torino: Bollati Boringhieri.

Friedman, Michael (1985). 'Kant's Theory of Geometry,' *Philosophical Review* 94: 455–506.

Friedman, Michael (1992). *Kant and the Exact Sciences,* Cambridge: Harvard University Press.

Friedman, Michael (2012). 'Kant on Geometry and Spatial Intuition,' *Synthese* 186: 231–55.

Galen (1964). *Institutio Logica,* English translation, introduction, and commentary by John Spangler Kieffer, Baltimore: The Johns Hopkins University Press.

Geach, Peter Thomas (1980 [1962]). *Reference and Generality: An Examination of Some Medieval and Modern Theories,* Ithaca/London: Cornell University Press.

Geach, Peter Thomas (1981). *Logic Matters,* Oxford: Basil Blackwell.

Griffel, Frank (2020). 'al-Ghazali,' *The Stanford Encyclopedia of Philosophy* (Summer 2020 Edition), ed. Edward N. Zalta. https://plato.stanford.edu/archives/sum2020/entries/al-ghazali/.

Grosseteste, Robert (1981). *Commentarius in Posteriorum Analyticorum libros,* ed. Pietro Rossi, Firenze: Leo Olschki Editore.

Hailperin, Theodore (2004). 'Algebraical Logic 1685–1900,' in Dov M. Gabbay and John Woods (eds.), *Handbook of the History of Logic,* vol. 3, *The Rise of Modern Logic from Leibniz to Frege,* Amsterdam/Boston/Tokyo: Elsevier/North Holland: 323–88.

Hanke, Miroslav (2020). 'Seventeenth Century Scholastic Syllogistic: Between Logic and Mathematics?,' *Review of Symbolic Logic* 13(2): 219–48.

Harari, Orna (2004). *Knowledge and Demonstration: Aristotle's Posterior Analytics,* The New Synthese Historical Library (Texts and Studies in the History of Philosophy), vol 56, Dordrecht: Springer.

Heath, Thomas L. (1949). *Mathematics in Aristotle,* Oxford: Clarendon Press.

Heath, Thomas L. (1956 [1925]). *The Thirteen Books of Euclid's Elements,* Cambridge: Cambridge University Press [reprinted in 1956 by Dover Publications, New York].

Heis, Jeremy (forthcoming). 'Did Kant Believe that Formal Logic is Analytic? Maybe Not' (draft 2018).

Henisch, Georg (1609). Arithmetica perfecta et demonstrata, doctrinam de numero triplici, vulgari, cossico & astronomico noua methodo per propositiones explicatam continens libris septem, Augustae Vindelicorum, [Augsburg]: David Franck.

Henninger, Mark (1989). *Relations: Medieval Theories 1250-1325,* Oxford: Oxford University Press.

Hérigone, Pierre (1634). *Cursus Mathematicus, nova, brevi, et clara methodo demonstratus per notas reales & universales, citra usum cuiuscunque idiomatis intellectu faciles—Cours Mathematique demonstré d'une nouvelle brève et Claire methode. Par notes reelles & universelles, qui peuvent estre entendues sans l'usage d'aucune langue*, Paris: chez l'Auteur et Henry Le Gras.

Herlinus, Christian and Dasypodius, Conrad (1566). *Analyseis Geometricae sex Librorum Euclidis. Primi et quinti factae a Christiano Herlino: reliquae una cum commentariis et scholiis perbrevibus in eosdem sex libros Geometricos a Conrado Dasypodio*, Strasbourg: Iosias Rihelius.

Hintikka, Jaakko (1967). 'Kant on the Mathematical Method,' *The Monist* 51(3): 352–75.

Hintikka, Jaakko (1969). 'On Kant's Notion of Intuition (Anschauung),' in Terrence Penelhum and John. J. MacIntosh (eds.), *The First Critique*, Belmont: Wadsworth Publishing Co.: 34–53.

Hintikka, Jaakko (1973). *Logic, Language-Games and Information*, Oxford: Clarendon Press.

Hintikka, Jaakko (1974). 'Kant's "new method of thought" and his theory of mathematics,' in Jaakko Hintikka, *Knowledge and the Known*, Dordrecht: D. Reidel Publishing Company: 126–34.

Hintikka, Jaakko (1978). 'Aristotle's Incontinent Logician,' *Ajatus* 37: 48–63.

Hintikka, Jaakko (1981). 'Kant's Theory of Mathematics Revisited,' *Philosophical Topics* 12 (2): 201–15.

Hintikka, Jaakko (1984). 'Kant's Transcendental Method and His Theory of Mathematics,' *Topoi* 3(2): 99–108.

Hintikka, Jaakko (1992). 'Kant on the Mathematical Method,' in Carl Posy (ed.), *Kant's Philosophy of Mathematics: Modern Essays*, Dordrecht: Kluwer Academic: 21–42.

Hintikka, Jaakko (2020). 'Kant's Theory of Mathematics: What Theory of What Mathematics?,' in Carl Posy and Ofra Rechter (eds.), *Kant's Philosophy of Mathematics: Volume I: The Critical Philosophy and Its Roots*, Cambridge: Cambridge University Press: 85–102.

Hintikka, Jaakko and Remes, Unto (1974). *The Method of Analysis: Its Geometrical Origin and Its General Significance*, Dordrecht: Reidel.

Hobbes, Thomas (1839). *Elements of Philosophy: The First Section, Concerning Body*, in *The English Works of Thomas Hobbes*, (ed.) William Molesworth, vol. 1. London: John Bohn.

Hodges, Wilfried (1998). 'The Law of Distribution for Syllogisms,' *Notre Dame Journal of Formal Logic* 39(2): 221–30.

Hoffmann, Adolf Friedrich (1729). *Gedancken über Christian Wolffens Logic*, Leipzig. Reprinted in Christian Wolff (2008), *Gesammelte Werke*, III.Abt., Bd. 117, Hildesheim: Georg Olms.

Hoffmann, Adolf Friedrich (1737). *Vernunftlehre*, Leipzig. Reprinted in two volumes in Christian Wolff (2010), *Gesammelte Werke*, III.Abt., Bd. 99.1 and 99.2, Hildesheim: George Olms.

Hogan, Desmond (2020). 'Kant and the Character of Mathematical Inference,' in Carl Posy and Ofra Rechter (eds.), *Kant's Philosophy of Mathematics: Volume I: The Critical Philosophy and Its Roots*, Cambridge: Cambridge University Press: 126–54.

Hurtado de Mendoza, Pedro (1619). *Disputationes in universam philosophiam a Summulis ad Metaphysicam, pars prior*. Mainz: ex typis et sumptibus Ioannis Albini.

Janssens, Jules (2020). 'al-Ghazālī's Maqāsˈid al-Falāsifa, Latin Translation of,' in Henrik Lagerlund (ed.), *Encyclopedia of Medieval Philosophy (Philosophy between 500 and 1500)*, Dordrecht: Springer: 630–3.

Jungius, Joachim (1957). *Logica Hamburgensis*, ed. R. W. Meyer (Joachim Jungius-Gesellschaft der Wissenschaften), Hamburg: in aedibus J. J. Augustin.

Kant, Immanuel (1992a). *Lectures on Logic*, trans. and ed. J. Michael Young, Cambridge: Cambridge University Press.

Kant, Immanuel (1992b). *Theoretical Philosophy 1755-1770*, trans. and ed. David Walford and Ralf Meerbote, Cambridge: Cambridge University Press.

Kant, Immanuel (1998). *Critique of Pure Reason*, trans. and ed. Paul Guyer and Allen W. Wood, Cambridge: Cambridge University Press.

Kant, Immanuel (2002). *Theoretical Philosophy after 1781*, ed. Henry Allison and Peter Heath, Cambridge: Cambridge University Press.

Kilwardby, Robert (2015). *Notule Libri Priorum*, critical edition with translation, introduction and indexes by Paul Thom and John Scott, 2 vols., Oxford: Oxford University Press.

Klügel, G. S. (1790). 'Recension. Versuch, der Einrichtung unsers Erkenntnisvermögens durch Algeber nachzuspüren. Von Chr. Ludw. Schübler. Leipzig 1788. 264 S. 8.,' in J. A. Eberhard, *Philosophisches Magazin*, Halle, J. Gebauer, vol. III: 236–49.

Kneale, William and Kneale, Martha (1962). *The Development of Logic*, Oxford: Clarendon Press.

Knuuttila, Simo (2010). 'Generality and Identity in Late Medieval Discussions of the Prior Analytics,' *Vivarium* 48: 215–27.

Krug, Wilhelm Traugott (1806). *System der theoretischen Philosophie. Erster Teil: Denklehre oder Logik*, Königsberg. Second edition 1819 (*Logik*).

Lambert, Johann Heinrich (1765). 'De Universaliori Calculi Idea, Disquisitio, una cum adnexo Specimine,' *Nova Acta Eruditorum* VI: 441–73.

Lambert, Johann Heinrich (1782). *Logische und philosophische Abhandlungen*, zum Druck befördert von Joh, Berlin: Bernoulli, erster Band.

Lambert, Johann Heinrich (1965). *Philosophische Schriften*, herausg. von Hans-Werner Arndt, Hildesheim: Georg Olms.

Lassalle-Casanave, Abel and Panza, Marco (2018). 'Enthymemathical Proofs and Canonical Proofs in Euclid's Plane Geometry,' in Hassan Tahiri (ed.), *The Philosophers and Mathematics: Festschrift for Roshdi Rashed*, Cham: Springer Verlag: 127–44.

Lear, Jonathan (1986). *Aristotle and Logical Theory*, Cambridge: Cambridge University Press.

Leibniz, Gottfried Wilhelm (1768). *Opera omnia, nunc primo collecta . . .* ed. L. Dutens, 6 vols., Geneva: apud Frates de Tournes.

Leibniz, Gottfried Wilhelm (1840). *Opera quae extant, Latina, Gallica, Germanica omnia*, ed. Johann E. Erdmann, Berlin: G. Eichler.

Leibniz, Gottfried Wilhelm (1965). *Die Philosophische Schriften*, ed. C. I. Gerhardt, 7 vols., Hildesheim: Olms Verlag [1st edn. Berlin, 1857–90].

Leibniz, Gottfried Wilhelm (1966). *Logical Papers*, trans. and ed. with an Introduction by G. H. R. Parkinson, Oxford: Oxford University Press.

Leibniz, Gottfried Wilhelm (1969). *Philosophical Papers and Letters*. A selection trans. and ed., with an Introduction by Leroy E. Loemker, Dordrecht: D. Reidel Publishing Company (2nd edn.).

Leibniz, Gottfried Wilhelm (1971). *Mathematische Schriften*, ed. C. I. Gerhardt, 7 vols., Hildesheim: Olms Verlag [1st edn. Halle, 1849–63].

Leibniz, Gottfried Wilhelm (1981). *New Essays on Human Understanding*, trans. and ed. P. Remnant and J. Bennett, Cambridge: Cambridge University Press.

Leibniz, Gottfried Wilhelm (1989). *Philosophical Essays*, ed. R. Ariew and D. Garber, Indianapolis/Cambridge: Hackett Publishing Company.

Leibniz, Gottfried Wilhelm (1999). *Sämtliche Schriften und Briefe. Philosophische Schriften*, herausgegeben von der Leibniz-Forschungsstelle der Universität Münster, Sechste Reihe, Band 4, Teilen A, B and C, Berlin: Akademie Verlag.

Leibniz, Gottfried Wilhelm (2020). *Dissertation on Combinatorial Art*, trans. and ed. M. Mugnai, H. van Ruler, and M. Wilson, Oxford: Oxford University Press.

Leichner, Johann Wilhelm Theodor (1727). *Die Entlarvete Und in ihrer wahren recht heßlichen Gestalt sich zeigende Rüdigersche Philosophie...*, Erfurt und Leipzig.

Lemke, Martin and Naujoks, Anne-Sophie (eds.) (2019). *Moritz Schlick. Vorlesungen und Aufzeichnungen zur Logik und Philosophie der Mathematik*, (Part of the Moritz Schlick Gesamtausgabe, Abteilung II: Nachgelassene Schriften Band 1. 3.), Wiesbaden: Springer.

Lenzen, Wolfgang (1984). 'Leibniz und die Boolesche Algebra,' *Studia Leibnitiana* 16 (2): 188–203.

Lenzen, Wolfgang (2004). 'Leibniz's Logic,' in Dov M. Gabbay and John Woods (eds.), *The Rise of Modern Logic: From Leibniz to Frege. Handbook of the History of Logic*, vol. 3. Amsterdam/Boston/ New York: Elsevier/North Holland: 1–83.

Lewis, Clarence Irving (1960). *A Survey of Symbolic Logic*, New York: Dover.

Locke, John (1997). *An Essay Concerning Human Understanding*, ed. Roger Woolhouse, Oxford: Penguin Classics.

Lohr, Charles H. (1965). 'Logica Algazelis: Introduction and Critical Text,' *Traditio* 21: 223–90.

Longuenesse, Béatrice (1998). *Kant and the Capacity to Judge: Sensibility and Discursivity in the Transcendental Analytic of the "Critique of Pure Reason,"* Princeton: Princeton University Press.

Lukasiewicz, Jan (1951). *Aristotle's Syllogistic from the Standpoint of Modern Formal Logic*, Oxford: Clarendon Press (2nd edn.).

Maaß, Johann Gebhard Ehrenreich (1793). *Grundriß der Logik*. Halle (3rd edn. 1806; 4th edn. 1823).

McKirahan, Richard (1992). *Principles and Proofs: Aristotle's Theory of Demonstrative Science*, Princeton: Princeton University Press.

Makinson, David (1969). 'Remarks on the Concept of Distribution in Traditional Logic,' *Noûs*, 3(1): 103–8.

Malink, Marko and Vasudevan, Anubav (2016). 'The Logic of Leibniz's Generales inquisitiones de analysi notionum et veritatum,' *The Review of Symbolic Logic* 9: 686–751.

Malink, Marko and Vasudevan, Anubav (2017). 'Leibniz's Theory of Propositional Terms: A Reply to Massimo Mugnai,' *The Leibniz Review* 27: 139–55.

Mancosu, Paolo (1996). *Philosophy of Mathematics and Mathematical Practice in the Seventeenth Century*, New York: Oxford University Press.

Mancosu, Paolo (2016). *Abstraction and Infinity*, Oxford: Oxford University Press.

Massa, Maria Rosa (2010). 'The Symbolic Treatment of Euclid's *Elements* in Hérigone's *Cursus mathematicus* (1634, 1637, 1642),' in A. Heeffer and M. Van Dyck (eds.), *Philosophical Aspects of Symbolical Reasoning in Early Modern Mathematics*, Vol. 26, London: College Publications: 165–91.

Mellado Romero, Antonio (2022). 'La Influencia del *Cursus Mathematicus* de Hérigone en la Algebrización de la Matemática,' Tesis de Doctorado, Universidad de Murcia.

Mendell, Henry (1998). 'Making Sense of Aristotelean Demonstration,' *Oxford Studies in Ancient Philosophy* XVI: 161–226.

Merrill, Daniel D. (1990). *Augustus De Morgan and the Logic of Relations*, Dordrecht/ Boston/London: Kluwer Academic.

Mill, John Stuart (1843). *System of Logic, Rationative and Inductive*. London: John Parker.

Mill, John Stuart (1974). *A System of Logic: Ratiocinative and Inductive*, Books I–III, ed. J. M. Robson, Introduction by F. R. McRae, Toronto: University of Toronto Press (volume VII of *The Collected Works of John Stuart Mill*, Liberty Press).

Minnema, Anthony H. (2014). 'Algazel Latinus: The Audience of the Summa Theoricae Philosophiae, 1150–1600,' *Traditio* 69: 153–213.

Mueller, Ian (1974). 'Greek Mathematics and Greek Logic,' in John Corcoran (ed.), *Ancient Logic and Its Modern Interpretations*, Dordrecht: Reidel: 35–70.

Mueller, Ian (1981). *Philosophy of Mathematics and Deductive Structure in Euclid's Elements*, Cambridge: The MIT Press.

Mugnai, Massimo (1992). *Leibniz's Theory of Relations*, Stuttgart: Steiner Verlag.

Mugnai, Massimo (2010). 'Logic and Mathematics in the Seventeenth Century,' *History and Philosophy of Logic* 31(4): 297–314.

Mugnai, Massimo (2016). 'Ontology and Logic: The Case of Scholastic and Late-Scholastic Theory of Relations,' *The British Journal for the History of Philosophy* 24(3): 532–51.

Müller, August Friedrich (1733). *Einleitung in die philosophischen Wissenschaften*, Leipzig (1st edn. 1728). Reprinted in Müller (2008), *Einleitung in die philosophischen Wissenschaften*, Mit einem Vorwort herausgegeben von Kay Zenker, Erster Teil, welcher den Eingang, die Logic, und Physic in sich enthält. I. Teilband, in Thomasiani. Materialen und Dokumente zu Christian Thomasius, herausgegeben von Werner Schneiders unter Mitarbeit von Kay Zenker, Band 3 Teil 1.1, Hildesheim/Zürich/New York: Georg Olms.

Murdoch, John E. (1969). '*Mathesis in philosophiam scholasticam introducta*: The Rise and Development of the Application of Mathematics in Fourteenth Century Philosophy and Theology,' in *Arts libéraux et philosophie au moyen âge*, Montréal/Paris: 215–56.

Murdoch, John E. (1978). 'The Development of a Critical Temper: New Approaches and Modes of Analysis in 14th Century Philosophy, Science and Theology,' *Medieval and Renaissance Studies* [Proceedings of the Southeastern Institute of Medieval and Renaissance Studies (Chapel Hill)] 7: 51–79.

Nuchelmans, Gabriel (1980). *Late-Scholastic and Humanist Theories of the Proposition*, Amsterdam/Oxford/New York: Verhandelingen der Kninklijke Nederlandse Akademie van Wetenschappen, Afd. Letterkunde, Nieuwe Reeks, Deel 103.

Nuchelmans, Gabriel (1983). *Judgment and Proposition: From Descartes to Kant*, Amsterdam/Oxford/New York: North-Holland.

Ockham, William of (1974). *Summa logicae*, in *Opera philosophica et theologica*, New York: St. Bonaventure, vol. I.

Ockham, William of (1997). *Ockham's Theory of Terms*, Part I of the *Summa logicae*, trans. and introduced by M. Loux, South Bend: St. Augustine's Press.

Ockham, William of (1998). *Ockham's Theory of Propositions*, Part II of the *Summa logicae*, trans. Alfred J. Freddoso and Henry Schuurman, introduced by J. Freddoso, South Bend: St. Augustine's Press.

Paccioni, Jean-Paul (2006). *Cet Esprit de Profondeur. Christian Wolff. L'Ontologie et la Métaphysique*, Paris: Vrin.

Pagli, Paolo (2009). 'Two Unnoticed Editions of Gerolamo Saccheri's *Logica Demonstrativa*,' *History and Philosophy of Logic* 30(4): 331–40.

Panteki, Maria (1991). 'Relationships between Algebra, Differential Equations and Logic in England, 1800–1860', PhD thesis, Middlesex University, London.

Panza, Marco (2018). Review of Thom, Paul and Scott, John, *Robert Kilwardby—Notule libri priorum. Part 1, 2. Critical edition with translation, introduction and indexes*. Dual Latin–English text. Published for The British Academy. Auctores Britannici Medii Aevi,

23–4, Oxford University Press, Oxford, 2015. cxi+1629, *Mathematical Reviews* 3675231, American Mathematical Society.

Paoli, Francesco (1990). 'Interpretazione di Alcune Teorie Logiche di Bolzano,' thesis, Università di Firenze.

Parsons, Charles (1992). 'Kant's Philosophy of Arithmetic,' in Carl J. Posy (ed.), *Kant's Philosophy of Mathematics. Modern Essays*, Dordrecht/Boston/London: Kluwer Academic: 135–58.

Parsons, Terence (2006). 'The Doctrine of Distribution,' *History and Philosophy of Logic* 27(1): 59–74.

Parsons, Terence (2008). *The Development of Supposition Theory in the Later 12th Through 14th Centuries*, in Dov E. Gabbay and John Woods (eds.), *Handbook of the History of Logic*, vol. 2, *Medieval and Renaissance Logic*, Amsterdam/Oxford: North-Holland: 157–280.

Parsons, Terence (2014). *Articulating Medieval Logic*, Oxford: Oxford University Press.

Pascal, Blaise (1995). *The Art of Persuasion*, in *Pensées and Other Writings*, trans. Honor Levi, ed. with introduction and notes by Anthony Levi, Oxford: Oxford University Press.

Patzig, Guenther (1968). *Aristotle's Theory of the Syllogism*, Dordrecht: D. Reidel.

Peckhaus, Volker (1997). *Logik, Mathesis universalis und allgemeine Wissenschaft. Leibniz und die Wiederentdeckung der formalen Logik im 19. Jahrhundert*, Berlin: Akademie Verlag.

Peckhaus, Volker (1999). 'Abduktion und Heuristik,' in J. Nida-Rümelin (ed.), *Rationalität, Realismus, Revision*, Berlin/New York: Walter de Gruyter: 833–41.

Pedrazzi, Maino (1974). 'Sul tentativo di Alessandro Piccolomini di ridurre a sillogismi la 1ᵃ dimostrazione degli *Elementi* di Euclide,' *Cultura e Scuola* 52: 221–30.

Peijnenburg, Jeanne (1994). 'Formal Proof or Linguistic Process? Beth and Hintikka on Kant's Use of "Analytic,"' *Kant-Studien* 85: 160–78.

Peirce, Charles Sanders (1931–58). *Collected Papers of Charles Sanders Peirce*, 8 vols., ed. Charles Hartshorne, Paul Weiss, and Arthur W. Burks, Cambridge: Harvard University Press (vols. 1–6 ed. Charles Harteshorne and Paul Weiss, 1931–5; vols. 7–8 ed. Arthur W. Burks, 1958).

Peirce, Charles Sanders (1982). *Writings of Charles S. Peirce: A Chronological Edition*, Volume I 1857–1866, ed. Peirce Edition Project, Bloomington: Indiana University Press.

Pelletier, Jacques [Iacobus Peletarius Cenomanus] (1557). *In Euclidis Elementa Geometrica Demonstrationum Libri sex*, Lyon: apud Ioan. Tornaesium et Gul. Gazeium.

Petrus, Klaus (1997). *Genese und Analyse. Logik, Rhetorik und Hermeneutik im 17. Und 18. Jahrhundert*, Berlin: Walter de Gruyter.

Philoponus, Iohannes (1905). *Philoponi in Analytica Priora Commentaria*, in *Commentaria in Aristotelem Graeca*, vol. XIII.2, ed. Maximilianus Wallies, Berlin: G. Reimer.

Philoponus, John (2008). *On Aristotle Posterior Analytics* 1.1–8, trans. Richard McKirahan, London: Bloomsbury.

Piccolomini, Alessandro (1547). *Alexandri Piccolominei In mechanicas quaestiones Aristotelis, paraphrasis paulo quidem plenior.* […] *Eiusdem Commentarium de certitudine mathematicarum disciplinarum*, Roma: Antonio Blado.

Proclus (1992 [1970]). *A Commentary on the First Book of Euclid's Elements*, trans. with Introduction and Notes by Glenn R. Morrow. With a new foreword by Ian Mueller, Princeton: Princeton University Press.

Proops, Ian (2005). 'Kant's Conception of Analytic Judgment,' *Philosophy and Phenomenological Research* LXX(3): 588–612.

Quintilianus, Marcus Fabius (1970). *Institutionis oratoriae libri duodecim*, recogn. brevique adnot. critica instruxit Michael Winterbottom, Oxonii: Clarendon Press.

Read, Stephen (2012). 'John Buridan's Theory of Consequence and His Octagons of Opposition,' in Jean-Yves Béziau and Dale Jacquette (eds.), *Around and Beyond the Square of Opposition*, Basel: Springer: 93–110.

Read, Stephen (2016). 'Non-Normal Propositions in Buridan's Logic,' in Laurent Cesalli et al. (eds.), *Formal Approaches and Natural Language in Medieval Logic*, Barcelona: Brepols: 453–68.

Reid, Thomas (1806). *Analysis of Aristotle's Logic*, 2nd edn., Edinburgh: William Creech.

Reid, Thomas (1822). *The Works of Thomas Reid, with an Account of his Life and Writings*, by Dougald Stewart, 3 vols., New York: J & J. Harper.

Reid, Thomas (1852). *The Works of Thomas Reid, D. D. now fully collected with selections from his unpublished letters. Preface, Notes and Supplementary Dissertations*, by Sir William Hamilton, Edinburgh: Maclachlan and Stewart.

Reyher, Samuel (1693). *De Euclide*, Kiliae.

Risse, Wilhelm (1970). *Die Logik der Neuzeit, 2. Band 1640–1780*, Stuttgart/Bad Cannstatt: Friedrich Frommann Verlag.

Ross, William David (1949). *Aristotle's Prior and Posterior Analytics. A Revised Text with Introduction and Commentary*, Oxford: Clarendon Press.

Rüdiger, Andreas (1707). *Philosophia Synthetica*, Lipsia. Reprinted in Rüdiger 2010b.

Rüdiger, Andreas (1709). *De Sensu Veri et Falsi*, Halle.

Rüdiger, Andreas (1711). *Institutiones Eruditionis*, Halle.

Rüdiger, Andreas (1716). *Physica Divina*, Frankfurt.

Rüdiger, Andreas (1717). *Institutiones Eruditionis*, Frankfurt.

Rüdiger, Andreas (1722). *De Sensu Veri et Falsi*, Leipzig.

Rüdiger, Andreas (1723). *Philosophia Pragmatica*, Leipzig. Reprinted in Rüdiger 2010a.

Rüdiger, Andreas (1727). *Christian Wolffens Meinung von dem Wesen der Seele und eines Geistes überhaupt, und Andreas Rüdiger Gegen-Meinung*, Leipzig. Reprinted in Christian Wolff (2008). *Gesammelte Werke*, III. Abt., Bd. 111, Hildesheim: Georg Olms.

Rüdiger, Andreas (1729). *Philosophia Pragmatica*, Leipzig.

Rüdiger, Andreas (2010a). *Philosophia Pragmatica*. Mit einem Vorwort herausgegeben von Ulrich Leinsle, I. Logica Physica, in *Thomasiani. Materialen und Dokumente zu Christian Thomasius*, herausgegeben von Werner Schneiders unter Mitarbeit von Kay Zenker, Band 5.2, Hildesheim/Zürich/New York: Georg Olms.

Rüdiger, Andreas (2010b). *Philosophia Synthetica*, Mit einem Vorwort herausgegeben von Ulrich Leinsle, in *Thomasiani. Materialen und Dokumente zu Christian Thomasius*, herausgegeben von Werner Schneiders unter Mitarbeit von Kay Zenker, Band 5.1, Hildesheim/Zürich/New York: Georg Olms.

Rusnock, Paul and Sebestik, Jan (2013). 'The Beyträge at 200: Bolzano's Quiet Revolution in the Philosophy of Mathematics,' *Journal for the History of Analytic Philosophy* 1(8): 1–14.

Russ, Steve (2004). *The Mathematical Works of Bernard Bolzano*, Oxford: Oxford University Press.

Russell, Bertrand (1903 [1938]). *The Principles of Mathematics*, Cambridge: Cambridge University Press.

Russell, Bertrand (1918). *Mysticism and Logic and Other Essays*, New York/London: Longmans, Green & Co.

Saccheri, Girolamo [Joannes Franciscus Caselette] (1697) [1980]. *Logica demonstrativa*, Augustae Taurinorum: Typis Ioannis Baptistae Zappatae [Hildesheim-New York: Olms].

Saccheri, Girolamo (1701). *Logica demonstrativa*, Ticini Regij: Caroli Francisci Magrij.

Saccheri, Girolamo (2012). *Logica dimostrativa*, ed. Massimo Mugnai and Massimo Girondino, Pisa: Edizioni della Normale.

Saccheri, Girolamo (2014). *Euclid Vindicated from Every Blemish*, ed. and annotated Vincenzo De Risi, trans. G. B. Halsted and L. Allegri, Basel: Birkhäuser.

Sánchez Valencia, Victor (1991). *Studies on Natural Logic and Categorical Grammar*, Amsterdam: University of Amsterdam Press.

Sánchez Valencia, Victor (1997). 'Head or Tail? De Morgan on the Bounds of Traditional Logic,' *History and Philosophy of Logic* 18: 123–38.

Sánchez Valencia, Victor (2004). 'The Algebra of Logic,' in Dov M. Gabbay and John Woods (eds.), *Handbook of the History of Logic*, vol. 3. *The Rise of Modern Logic from Leibniz to Frege*, Amsterdam/Boston/Tokyo: Elsevier/North Holland: 389–544.

Savonarola, Girolamo (1982). *Compendium Logicae*, in *Scritti filosofici*, a cura di Giancarlo Garfagnini e Eugenio Garin, Roma: Angelo Belardetti, vol. I.

Scheibler, Christoph (1654). *Opus Logicum, Quatuor Partibus, Universum Hujus Artis Systema Comprehendens...*, Giessae Hassorum: Chemlin.

Schepers, Heinrich (1959). *Andreas Rüdigers Methodologie und ihre Voraussetzungen*, Köln: Kölner Universitäts-Verlag.

Schleiermacher, Friederich (1839). *Dialektik. Aus Schleiermachers handschriftlichen Nachlasse*, hg. von L. Jonas, Berlin: Reimer (also in Friedrich Schleiermacher's Sämmtliche Werke, 3.Abt.: Zur Philosophie, Bd. 4.2).

Schleiermacher, Friederich (2002). 'Vorlesungen über Dialektik,' in F. Schleiermacher, *Kritische Gesamtausgabe*, ed. Andreas Arndt, Band 10/Teilband 1+2 Vorlesungen über die Dialektik, Berlin/Boston: de Gruyter.

Schlick, Moritz (1918 [1925]). *Allgemeine Erkenntnislehre*, Berlin: Julius Springer. English translation of the 2nd edn. in Schlick (1985).

Schlick, Moritz (1985). *General Theory of Knowledge*, Chicago: Open Court.

Schneider, Martin (1988). 'Funktion und Grundlegung der *Mathesis Universalis* in Leibnizschen Wissenschaftsystem,' in Albert Heinekamp (ed.), *Leibniz: questions de logique*, Studia Leibnitiana, Sonderheft 15: 162–82.

Schuman, Boaz Faraday (2022). 'Multiple Generality in Scholastic Logic,' *Oxford Studies in Medieval Philosophy*, 10: 215–82.

Sgarbi, Marco and Cosci, Matteo (eds.) (2018). *The Aftermath of Syllogism: Aristotelian Logical Argument from Avicenna to Hegel*, London: Bloomsbury.

Sigwart, Christoph (1895). *Logic*, London: Swann Sonnenschein & Co.

Smiley, Timothy J. (1973). 'What Is a Syllogism?,' *Journal of Philosophical Logic* 2(1): 136–54.

Smith, Robin (1982). 'What Is Aristotelian *ecthesis*?,' *History and Philosophy of Logic* 3: 113–27.

Stelling, Jendrich (ed.) (2019). *Moritz Schlick. Frühe erkenntnistheoretische Schriften*, Springer. Part of the Moritz Schlick Gesamtausgabe, Abteilung II: Nachgelassene Schriften Band 1.1, Wiesbaden: Springer.

Strobino, Riccardo (2021). *Avicenna's Theory of Science: Logic, Metaphysics, Epistemology*, Berkeley: University of California Press.

Sturm, Johann Christopher (1661). *Joh. Christophori Sturmii Universalia Euclidea*, Hagae-Comitis: Ulacq.

Sylla, Edith S. (1973). 'Medieval Concepts of the Latitude of Forms: The Oxford Calculators,' *Archives d'histoire doctrinale et littéraire du Moyen Age* 40: 223–83.

Tarski, Alfred (1941). 'On the Calculus of Relations,' *The Journal of Symbolic Logic* 6(3): 73–89.

Thom, Paul (1976). 'Ecthesis,' *Logique et Analyse* 74: 299–310.

Thom, Paul (1977). '*Termini Obliqui* and the Logic of Relations,' *Archiv für Geschichte der Philosophie* 59: 143–55.

Thom, Paul (1981). *The Syllogism*, Munich: Philosophia Verlag.

Thom, Paul (2016). *The Syllogism and Its Transformations*, in Dutilh Novaes and Read 2016: 290–315.

Thomas of Aquin (1980). *S. Thomae Opera Omnia*, Stuttgart/Bad Cannstadt: Steiner Verlag.

Toletus, Franciscus (1579). *Commentaria cum quaestionibus in universam Aristotelis logicam*, Lugduni.

Ueberweg, Friedrich (1857). *System der Logik und Geschichte der logischen Lehren*, Bonn: Adulph Marcus. Later editions: 1865²; 1868³; 1874⁴; 1882⁵. Translated as *System of Logic and History of Logical Doctrines*, London: Longmans, Green, and Co., 1871.

Vagetius, Johannes (1977). *Logicae Hamburgensis Additamenta. Cum annotationibus edidit Wilhelm Risse*, Göttingen: Vandenhoeck & Ruprecht.

Vailati, Giovanni (1903). 'Di un'opera dimenticata del P. Gerolamo Saccheri (Logica demonstrativa, *1697*),' *Rivista filosofica* 1903: 212–19.

Van Benthem, Johann (1974). 'Hintikka on Analyticity,' *Journal of Philosophical Logic* 3: 419–31.

Van Benthem, Johann (1986). *Essays in Logical Semantics*, Dordrecht: D. Reidel.

Van Benthem, Johann and ter Meulen, Alice (eds.) (1985). *Generalized Quantifiers in Natural Language*, Dordrecht: Foris.

Van Eijck, Dingeman Johannes Norbertus (1985). 'Aspects of Quantification in Natural Language,' dissertation, Rijksuniversiteit, Groningen.

Varro, Terentius Marcus (1910). *De lingua Latina quae supersunt*, recensuerunt Georgius Goetz et Fridericus Schoell, Leipzig: Teubner.

Weigel, Erhard (1658). *Analysis Aristotelica ex Euclide restituta*, Jena: Grosium. Reprinted in Weigel (2008).

Weigel, Erhard (1693). *Philosophia Mathematica, Theologia naturalis, solida per singula scientias continuata universae artis inveniendi prima stamina complectens*, Jena: M. Birkner.

Weigel, Ehrard (2008 [1658]). *Analysis Aristotelica ex Euclide restituta*, in E. Weigel, *Werke III*, herausgegeben und eingeleitet von Thomas Behme, Stuttgart/Bad Cannstatt: Frommann-Holzboog.

William, Bishop of Lucca (1975). *Summa Dialetice Artis*, ed. Lorenzo Pozzi, Padova: Liviana Editrice.

Wolff, Christian (1704). *Dissertatio algebraica de algorithmo infinitesimali differentiali*, Lipsiae. Reprinted in C. Wolff (1974), *Gesammelte Werke*, II.Abt., Band 35, *Melemata mathematico-philosophica*, Hildesheim: Georg Olms.

Wolff, Christian (1709). *Aëreometriae Elementa*, Lipsiae. Reprinted in C. Wolff (2001), *Gesammelte Werke*, II.Abt., Band 37, Hildesheim: Georg Olms.

Wolff, Christian (1710). *Der Anfangs-Gründe aller mathematischen Wissenschaften*, Halle. 2nd edn. 1750. 1750 edition reprinted in C. Wolff (1973). *Gesammelte Werke*, I.Abt., 12–15.2, Hildesheim, Georg Olms. Latin translation in Wolff (1713–15).

Wolff, Christian (1713). *Vernünfftige Gedancken von den Kräfften des menschlichen Verstandes* (*German Logic*). 1754 edn. reprinted in C. Wolff (1965), *Gesammelte Werke*, I.Abt., Band 1, Hildesheim: Georg Olms.

Wolff, Christian (1713–15). *Elementa matheseos universae*, Halle. Latin translation of Wolff (1710).

Wolff, Christian (1718). *Ratio praelectionum Wolfianarum in Mathesin et Philosophiam universam*, Halle. 1735 edn. reprinted in C. Wolff (1972), *Gesammelte Werke*, II.Abt., Band 36, Hildesheim: Georg Olms.

Wolff, Christian (1720). *Vernünfftige Gedancken vom Gott, der Welt un der Seele des Menschen* [German metaphysics], Halle. 1751 edn. reprinted in C. Wolff (1983), *Gesammelte Werke*, I.Abt., Band 2 (1, 2), Hildesheim: Georg Olms.

Wolff, Christian (1728). *Philosophia Rationalis Sive Logica, methodo scientifica pertractata* [Latin Logic], Francofurti et Lipsiae. 1740 edn. reprinted in C. Wolff (1983), *Gesammelte Werke*, II.Abt., Band 1 (1–3), Hildesheim: Georg Olms.

Wolff, Christian (1730). *Philosophia prima sive Ontologia, methodo scientifica pertractata* [Latin metaphysics], Francofurti et Lipsiae. 1736 edn. reprinted in C. Wolff (1962), *Gesammelte Werke*, II.Abt., Band 3, Hildesheim: Georg Olms.

Wolff, Christian (1731). *Kurtzer Unterricht von den vornehmsten mathematischen Schrifften*, Franckfurt und Leipzig. 1750 edn. reprinted in C. Wolff (1973), *Gesammelte Werke*, I.Abt., Band 15 (2), Hildesheim: Georg Olms.

Wolff, Christian (1732). *Psychologia empirica, methodo scientifica pertractata*, Francofurti et Lipsiae. 1738 edn. reprinted in C. Wolff (1968), *Gesammelte Werke*, II.Abt., Band 5, Hildesheim: Georg Olms.

Wolff, Christian (1735). *Philosophia Rationalis sive Logica, Methodo scientifica pertractata*, Veronae: ex typographia Dionysii Ramanzini.

Wolff, Christian (1841). *Eigene Lebenschreibung, mit einer Abhandlung über Wolff*, ed. von H. Wuttke, Leipzig. Reprinted in C. Wolff (1980), *Gesammelte Werke*, I.Abt., Band 10, Hildesheim: Georg Olms.

Wolff, Christian (1860). *Briefwechsel zwischen Leibniz und Christian Wolff*, ed. C. J. Gerhardt, Halle (reprinted by Georg Olms, Hildesheim, 1963).

Wundt, Wilhelm (1945). *Die Deutsche Schulphilosophie im Zeitalter der Aufklärung*, Tübingen: Mohr (reprinted by Georg Olms Verlag, Hildesheim, 1992).

Young, J. Michael (1992). 'Translator's Introduction,' in I. Kant, *Lectures on Logic*, trans. and ed. J. Michael Young, Cambridge: Cambridge University Press, xv–xxxii.

Index of Names

For the benefit of digital users, indexed terms that span two pages (e.g., 52–53) may, on occasion, appear on only one of those pages.